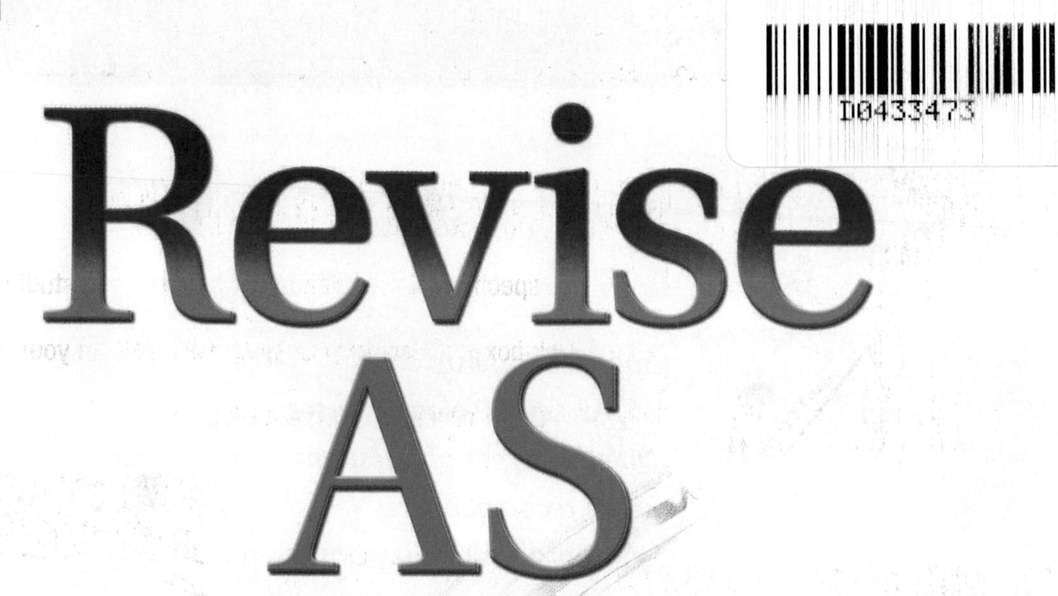

Revise AS Biology

John Parker & Ian Honeysett

Contents

Chapter 3 Enzymes

Chapter 4 Exchange

Chapter 5 Transport

Contents

Specification lists

AQA Biology

MODULE	SPECIFICATION TOPIC	CHAPTER REFERENCE	STUDIED IN CLASS	REVISED	PRACTICE QUESTIONS
Unit 1	Disease and lifestyle	8.1, 8.2			
	Digestion	3.3			
	Biological molecules	1.1,1.2,1.3,1.4,1.6			
	Enzymes	3.1, 3.2, 3.4			
	Cells and membrane transport	2.1, 2.2, 2.3, 4.1, 4.2, 8.2			
	The lungs and lung disease	4.3, 8.1			
	The heart and heart disease	5.2, 8.1			
	Immunity	8.3			
Unit 2	Variation	7.2			
	DNA and the genetic code	6.1			
	Cell division	6.2			
	Biochemical differences	2.1, 1.2, 5.4			
	Cell differentiation	2.3			
	Transport	5.1, 5.3, 5.5			
	Gaseous exchange	4.4, 5.5			
	Classification	7.2			
	Mutation and variation	7.2			
	Biodiversity	7.2			

Examination analysis

Unit 1 *– Examination paper (100 UMS) (60 raw marks) 5–7 short answer questions plus 2 longer questions, a short comprehension and a short structured essay 1 1/4 hours 33.33% of the total AS marks 16.67% of the total A-level marks*

Unit 2 *– Examination paper (140 UMS) (85 raw marks) 7–9 short answer questions plus 2 longer questions (1 data handling and 1 assessing analysis and evaluation) 1 3/4 hours 46.67% of the total AS marks 23.33% of the total A-level marks*

Unit 3 – Investigative and practical skills in AS Biology *AS Centre Assessed Unit (60 UMS) (50 raw marks) Practical Skills Assessment (PSA) 6 marks Investigative Skills Assignment (ISA) 44 marks 20% of the total AS marks 10% of the total A-level marks*

OCR Biology

MODULE	SPECIFICATION TOPIC	CHAPTER REFERENCE	STUDIED IN CLASS	REVISED	PRACTICE QUESTIONS
Unit 1 Module 1	*Cell structure*	2.1, 2.2, 2.3			
	Cell membranes	4.1, 4.2			
	Cell division	6.2			
	Cell differentiation	2.3			
Unit 1 Module 2	*Gaseous exchange*	4.3, 4.4			
	Transport in animals	5.1, 5.2, 5.3, 5.4			
	Transport in plants	5.5			
Unit 2 Module 1	*Biological molecules*	1.1, 1.2, 1.3, 1.4, 1.5, 1.6			
	Nucleic acids	6.1			
	Enzymes	3.1, 3.2			
Unit 2 Module 2	*Diet and Food*	8.1			
	Disease and immunity	8.1, 8.2, 8.3			
Unit 2 Module 3	*Biodiversity*	7.2			
	Classification	7.1			
	Evolution	7.2			
	Maintaining biodiversity	7.3			

Examination analysis

Unit 1	*1 hour written exam* *AS Level – 30%* *A Level – 15%*
Unit 2	*1 hour 45 mins written exam* *AS Level – 50%* *A Level – 25%*
Unit 3	*Internal assessment* *AS Level – 20%* *A Level – 10%*

WJEC Biology

MODULE	SPECIFICATION TOPIC	CHAPTER REFERENCE	STUDIED IN CLASS	REVISED	PRACTICE QUESTIONS
Unit 1	*Biological molecules*	*1.1, 1.2, 1.3, 1.4, 1.5, 1.6*			
	Cell structure and organisation	*2.1, 2.2, 2.3*			
	Cell membranes and transport	*4.1, 4.2*			
	Enzymes	*3.1, 3.2, 3.3, 3.4*			
	Nucleic acids	*6.1*			
	Cell division	*6.2*			
Unit 2	*Biodiversity*	*7.2*			
	Classification	*7.1*			
	Gas exchange	*4.3, 4.4*			
	Transport in animals	*5.2, 5.3, 5.4*			
	Transport in plants	*5.5*			
	Reproductive strategies				
	Adaptations for nutrition	*3.3*			

Examination analysis

Biology 1 *20% 1 hour 30 min Written Paper 70 marks (120UM)*
Unit title *Basic Biochemistry and cell organisation*
Outline of paper structure
– Short and longer structured questions, choice of 1 from 2 essays.

Biology 2 *20% 1 hour 30 min Written Paper 70 marks (120UM)*
Unit Title *Biodiversity and physiology of Body Systems*
Outline of paper structure
– Short and longer structured questions, choice of 1 from 2 essays.

Biology 3 *10% Internal assessment 44 marks (60UM)*
AS Unit *AS Practical assessment*
Experimental work set in centre, completed by candidates over 3 month period. Marked by board plus, low power plan microscope drawing.

CCEA Biology

MODULE	SPECIFICATION TOPIC	CHAPTER REFERENCE	STUDIED IN CLASS	REVISED	PRACTICE QUESTIONS
Unit 1 (M1)	*Molecules*	1.1, 1.2, 1.3, 1.4, 1.5,1.6			
	Enzymes	3.1, 3.2, 3.3, 3.4			
	DNA Technology	6.3			
	Viruses	8.2			
	Cells	2.1, 2.2, 4.1			
	Cell physiology	4.2,			
	Continuity of cells	6.2			
	Tissues and organs	2.3, 3.3, 5.5			
Unit 2 (M2)	*Principles of exchange*	5.1			
	Gaseous exchange	4.3,			
	Transport in plants	5.5			
	Transport in mammals	5.2, 5.3, 5.4, 8.1			
	Adaptations of organisms	7.2			
	Biodiversity and Classification	7.1, 7.2			
	Human impact	7.3			

Examination analysis

AS 1: Molecules and Cells
1 hour 30 minutes written examination, externally assessed
40% of AS 20% of A Level
January and Summer

AS 2: Organisms and Biodiversity
1 hour 30 minutes written examination, externally assessed
40% of AS 20% of A Level
January and Summer

AS 3: Assessment of Practical Skills in AS Biology
Internal practical assessment
20% of AS 10% of A Level
Summer only

Edexcel Biology

MODULE	SPECIFICATION TOPIC	CHAPTER REFERENCE	STUDIED IN CLASS	REVISED	PRACTICE QUESTIONS
Unit 1	Biological molecules	1.1, 1.2, 1.3, 1.4			
	Cell membranes and transport	4.1, 4.2			
	Transport in animals	5.2, 5.3, 8.1			
	DNA and protein synthesis	6.1			
	Inheritance				
	Mutations	7.2			
Unit 2B	Cell structure	2.1, 2.2, 2.3			
	Cell division	6.2			
	Variation	7.2			
	Biodiversity	7.2			
	Classification	7.1			
	Transport in plants	5.5			

Examination analysis

Unit 1 AS
This unit is assessed by means of a written examination paper, which lasts 1 hour 15 minutes and will include: objective questions structured questions short-answer questions and will also cover: How Science Works practical-related questions. 80 marks

Unit 2 AS
This unit is assessed by means of a written examination paper, which lasts 1 hour 15 minutes and will include: objective questions structured questions short-answer questions and will also cover: How Science Works practical-related questions. 80 marks

Unit 3 AS
Students will submit a written report of between 1500 and 2000 words which will be marked by the teacher and moderated by Edexcel. The report may be either a record of a visit to a site of biological interest or a report of research into a biological topic. During the course teachers will observe students carrying out practical work and submit their assessment (non-moderated). There is no separate content for this unit. 50 marks

AS/A2 Level Biology courses

AS and A2

All Biology A Level courses being studied from September 2008 are in two parts, with a number of separate modules or units in each part. Most students will start by studying the AS (Advanced Subsidiary) course. Some will go on to study the second part of the A Level course, called A2. It is also possible to study the full A Level course in either order. Advanced Subsidiary is assessed at the standard expected halfway through an A Level course, i.e. between GCSE and A Level. This means that the new AS and A2 courses are designed so that difficulty steadily increases:

- AS Biology builds from GCSE Science and Additional Science/Biology.
- A2 Biology builds from AS Biology.

How will you be tested?

Assessment units

AS Biology comprises three units or modules. The first two units are assessed by examinations. The third component usually involves some method of practical assessment (this is dependent on the Examination Group).

Centre-based coursework involves practical skills marked by your teacher. The marks can be adjusted by moderators appointed by the awarding body.

For AS Biology, you will be tested by three assessment units. For the full A Level in Biology, you will take a further three units. AS Biology forms 50% of the assessment weighting for the full A Level.

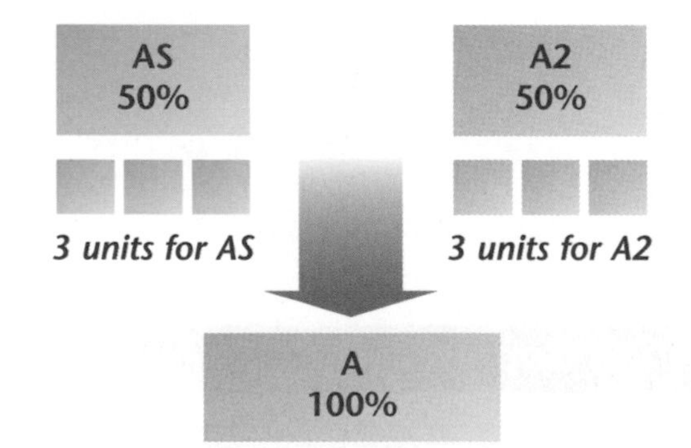

Tests are taken at two specific times of the year, January and June. It can be an advantage to you to take a unit test at the earlier optional time because you can re-sit the test, **(only once!)**. The best mark from the two will be credited and the lower mark ignored.

Each unit can normally be taken in either January or June. Alternatively, you can study the whole course before taking any of the unit tests. There is a lot of flexibility about when exams can be taken and the diagram below shows just some of the ways that the assessment units may be taken for AS and A Level Biology.

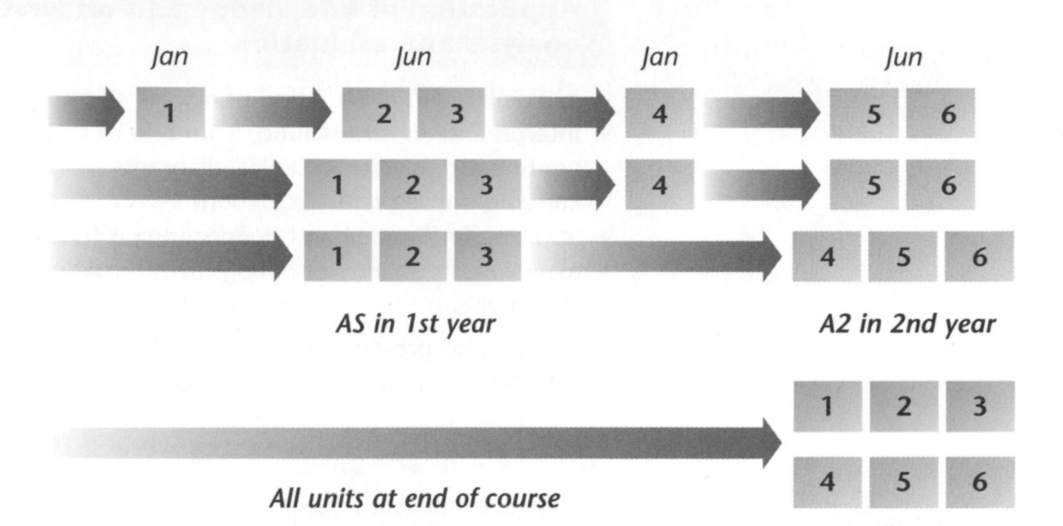

If you are disappointed with a module result, you can re-sit each module once. You will need to be very careful about when you take up a re-sit opportunity because you will have only one chance to improve your mark. The higher mark counts.

A2 and synoptic assessment

Many students who have studied at AS Level may decide to go on to study A2. There are three further units or modules to be studied. Every A Level specification includes a 'synoptic' assessment at the end of A2. Synoptic questions make use of concepts from earlier units, bringing them together in holistic contexts. Examiners will test your ability to inter-relate topics through the complete course from AS to A2.

Coursework

Coursework may form part of your A Level Biology course, depending on which specification you study. Where students have to undertake coursework, it is usually for the assessment of practical skills but this is not always the case.

What skills will I need?

For AS Biology, you will be tested by assessment objectives: these are the skills and abilities that you should have acquired by studying the course. The assessment objectives for AS Biology are shown below.

Knowledge with understanding

- recall of facts, terminology and relationships
- understanding of principles and concepts
- drawing on existing knowledge to show understanding of the responsible use of biological applications in society
- selecting, organising and presenting information clearly and logically

Application of knowledge and understanding, analysis and evaluation

- explaining and interpreting principles and concepts
- interpreting and translating, from one to another, data presented as continuous prose or in tables, diagrams and graphs
- carrying out relevant calculations
- applying knowledge and understanding to familiar and unfamiliar situations
- assessing the validity of biological information, experiments, inferences and statements

You must also present arguments and ideas clearly and logically, using specialist vocabulary where appropriate. Remember to balance your argument!

Experimental and investigative skills

Biology is a practical subject and part of the assessment of AS Biology will test your practical skills. This may be done during your lessons or may be tested in a more formal practical examination. You will be tested on four main skills:

- planning
- implementing
- analysing evidence and drawing conclusions
- evaluating evidence and procedures.

The skills may be assessed in the context of separate practical exercises, although more than one skill may be assessed in any one exercise. They may also be assessed all together in the context of a 'whole investigation'. An investigation may be set by your teacher or you may be able to pursue an investigation of your own choice.

You will receive guidance about how your practical skills will be assessed from your teacher. This study guide concentrates on preparing you for the written examinations testing the subject content of AS Biology.

Different types of questions in AS examinations

In AS Biology examinations different types of questions are used to assess your abilities and skills. Unit tests mainly use structured questions requiring both short-answers and more extended answers.

Short-answer questions

A question will normally begin with a brief amount of stimulus material. This may be in the form of a diagram, data or graph. A short-answer question may begin by testing recall. Usually this is followed up by questions which test understanding. Often you will be required to analyse data.

Short-answer questions normally have a space for your responses on the printed paper. The number of lines is a guide as to the amount of words you will need to answer the question. The number of marks indicated on the right side of the paper shows the number of marks you can score for each question part.

Here are some examples. (The answers are shown in blue)

The diagram shows part of a DNA molecule.

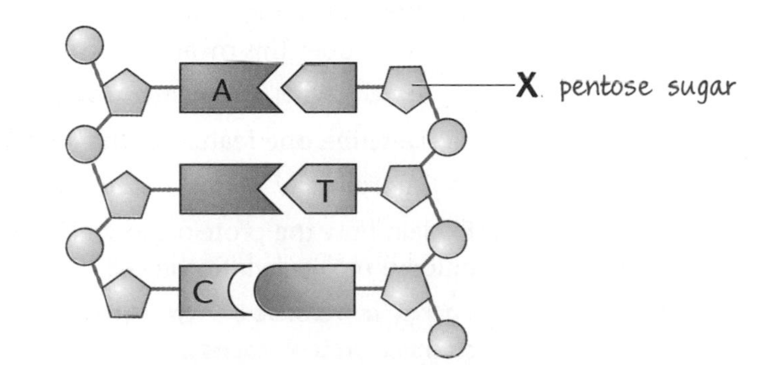

(a) Label part X. [1]

(b) Complete the diagram by writing a letter for each missing organic base in each empty box. [1]

(b) How do two strands of DNA join to each other?

The organic bases ✓ link the strands by hydrogen bonds ✓ [2]

Structured questions

Structured questions are in several parts. The parts are usually about a common context and they often progress in difficulty as you work through each of the parts. They may start with simple recall, then test understanding of a familiar or unfamiliar situation. If the context seems unfamiliar the material will still be centred around concepts and skills from the Biology specification. (If a student can answer questions about unfamiliar situations then they display understanding rather than simple recall.)

The most difficult part of a structured question is usually at the end. Ascending in difficulty, a question allows a candidate to build in confidence. Right at the end technological and social applications of biological principles give a more demanding challenge. Most of the questions in this book are structured questions. This is the main type of question used in the assessment of AS Biology.

When answering structured questions, do not feel that you have to complete a question before starting the next. Answering a part that you are sure of will build your confidence. If you run out of ideas go on to the next question. This will be more profitable than staying with a very difficult question which slows down progress. Return at the end when you have more time.

Here is an example of a structured question which becomes progressively more demanding.

Question

The diagram shows the molecules of a cell surface membrane.

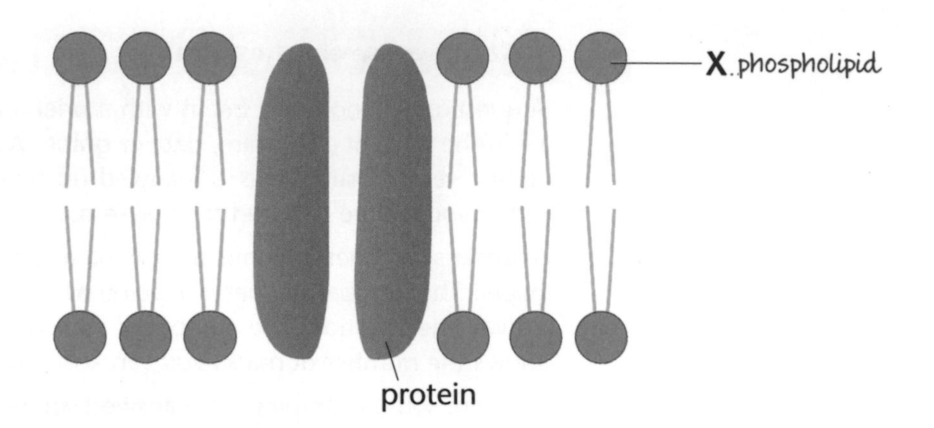

(a) (i) Label molecule X. [1]

(ii) The part of molecule X facing the outside of a cell is hydrophilic. What does this mean?

water loving/water attracting [1]

(iii) Describe **one** feature of the part of molecule X which faces inwards.

Hydrophobic/ water hating fatty acid residues [1]

(b) Explain how the protein shown in the diagram can actively transport the glucose molecule into the cell.

Energy is released from mitochondria near the channel protein the channel protein opens [3]

Note the help given in diagrams. The labelling of the protein molecule may trigger the memory so that the candidate has to make a small step to link the 'channel' function to this diagram. Examiners give clues! Expect more clues at AS Level than at A2 Level.

Extended answers

In AS Biology, questions requiring more extended answers will usually form part of structured questions. They will normally appear at the end of a structured question and will typically have a value of three to six marks. Longer answers are allocated more lines, so you can use this as a guide as to the extent of your answer. The mark allocation is a guide as to how many points you need to make in your response. Often for an answer worth six marks the mark scheme could have eight creditable answers. You are awarded up to the maximum, six in this instance.

Depending on the awarding body, longer extended questions may be set. These are often open response questions. These questions may be worth up to ten marks for full credit. Extended answers are used to allocate marks for the **quality of written communication**.

Candidates are assessed on their ability to use a suitable style of writing, and organise relevant material, both logically and clearly. The use of specialist biological words in context is also assessed. Spelling, punctuation and grammar are also taken into consideration. Here is a longer extended response question.

Question

Give an account of the effects of sewage entry into a river and explain the possible consequences to organisms downstream.

The sewage enters the river and is decomposed by bacteria. ✓ These bacteria are saprobiotic ✓ they produce nitrates which act as a fertiliser. ✓ Algae form a blanket on the surface ✓ light cannot reach plants under the algae so these plants die. ✓ Bacteria decompose the dead plants ✓ the bacteria use oxygen/bacteria are aerobic ✓ fish die due to lack of oxygen ✓ Tubifex worms or bloodworms increase near sewage entry ✓ mayfly larvae cannot live close to sewage entry/mayfly larvae appear a distance downstream where oxygen levels return. ✓

10 marking points → [7]

Remember that mark schemes for extended questions often exceed the question total, but you can only be awarded credit up to a maximum. Examiners sometimes build in a hurdle, e.g. in the above responses, references to one organism which increases in population is worthy of a mark, and another which decreases in population is worth another. Continually referring to different species which repeat a growth pattern will not gain further credit.

Exam technique

Links from GCSE

AS Biology builds from grade C in GCSE Science and GCSE Additional Science (combined) or GCSE Biology. This study guide has been written so that you will be able to tackle AS Biology from a GCSE science background.

You should not need to search for important Biology from GCSE science because this has been included where needed in each chapter. If you have not studied science for some time, you should still be able to learn AS Biology using this text alone.

What are examiners looking for?

Whatever type of question you are answering, it is important to respond in a suitable way. Examiners use instructions to help you to decide the length and depth of your answer. The most common words used are given below, together with a brief description of what each word is asking for.

Define

This requires a formal statement. Some definitions are easy to recall.

Define the term active transport.

This is the movement of molecules from where they are in lower concentration to where they are in higher concentration. The process requires energy.

Other definitions are more complex. Where you have problems it is helpful to give an example.

Define the term endemic.

This means that a disease is found regularly in a group of people, district or country. Use of an example clarifies the meaning. Indicating that malaria is invariably found everywhere in a country, confirms understanding.

Explain

This requires a reason. The amount of detail needed is shown by the number of marks allocated.

Explain the difference between resolution and magnification.

Resolution is the ability to be able to distinguish between two points whereas magnification is the number of times an image is bigger than an object itself.

State

This requires a brief answer without any reason.

State one role of blood plasma in a mammal.

Transport of hormones to their target organs.

List

This requires a sequence of points with no explanation.

List the abiotic factors which can affect the rate of photosynthesis in pond weed.

carbon dioxide concentration; amount of light; temperature; pH of water

Describe

This requires a piece of prose which gives key points. Diagrams should be used where possible.

Describe the nervous control of heart rate.

The medulla oblongata ✓ of the brain connects to the sino atrial node in the right atrium, wall ✓ via the vagus nerve and the sympathetic nerve ✓ the sympathetic nerve speeds up the rate ✓ the vagus nerve slows it down.✓

Discuss

This requires points both for and against, together with a criticism of each point. (**Compare** is a similar command word.)

Discuss the advantages and disadvantages of using systemic insecticides in agriculture.

Advantages are that the insecticides kill the pests which reduce yield ✓ they enter the sap of the plants so insects which consume sap die ✓ the insecticide lasts longer than a contact insecticide, 2 weeks is not uncommon ✓

Disadvantages are that insecticide may remain in the product and harm a consumer e.g. humans ✓ it may destroy organisms other than the target ✓ no insecticide is 100% effective and develops resistant pests. ✓

Suggest

This means that there is no single correct answer. Often you are given an unfamiliar situation to analyse. The examiners hope for logical deductions from the data given and that, usually, you apply your knowledge of biological concepts and principles.

The graph shows that the population of lynx decreased in 1980. Suggest reasons for this.

Weather conditions prevented plant growth ✓ so the snowshoe hares could not get enough food and their population remained low ✓ so the lynx did not have enough hares(prey) to predate upon. ✓ The lynx could have had a disease which reduced numbers. ✓

Calculate

This requires that you work out a numerical answer. Remember to give the units and to show your working, marks are usually available for a partially correct answer. If you work everything out in stages write down the sequence. Otherwise if you merely give the answer and it is wrong, then the working marks are not available to you.

Calculate the Rf value of spot X. (X is 25 mm from start and solvent front is 100 mm)

$$Rf = \frac{\text{distance moved by spot}}{\text{distance moved by the solvent front}}$$

$$= \frac{25 \text{ mm}}{100 \text{ mm}}$$

$$= 0.25$$

Outline

This requires that you give only the main points. The marks allocated will guide you on the number of points which you need to make.

Outline the use of restriction endonuclease in genetic engineering.

The enzyme is used to cut the DNA of the donor cell. ✓

It cuts the DNA up like this AT|GCCGAT = AT + GCCGAT ✓
TACGGC|TA TACGGC TA

The DNA in a bacterial plasmid is cut with the same restriction endonuclease. ✓

The donor DNA will fit onto the sticky ends of the broken plasmid. ✓

If a question does not seem to make sense, you may have mis-read it. Read it again!

Some dos and don'ts

Dos

Do answer the question

No credit can be given for good Biology that is irrelevant to the question.

Do use the mark allocation to guide how much you write

Two marks are awarded for two valid points – writing more will rarely gain more credit and could mean wasted time or even contradicting earlier valid points.

Do use diagrams, equations and tables in your responses

Even in 'essay style' questions, these offer an excellent way of communicating biology.

Do write legibly

An examiner cannot give marks if the answer cannot be read.

Do write using correct spelling and grammar. Structure longer essays carefully

Marks are now awarded for the quality of your language in exams.

Don'ts

Don't fill up any blank space on a paper

In structured questions, the number of dotted lines should guide the length of your answer.

If you write too much, you waste time and may not finish the exam paper. You also risk contradicting yourself.

Don't write out the question again

This wastes time. The marks are for the answer!

Don't contradict yourself

The examiner cannot be expected to choose which answer is intended. You could lose a hard-earned mark.

Don't spend too much time on a part that you find difficult

You may not have enough time to complete the exam. You can always return to a difficult calculation if you have time at the end of the exam.

What grade do you want?

Everyone would like to improve their grades but you will only manage this with a lot of hard work and determination. You should have a fair idea of your natural ability and likely grade in biology and the hints below offer advice on improving that grade.

For a Grade A

You will need to be a very good all-rounder.

- You must go into every exam knowing the work extremely well.
- You must be able to apply your knowledge to new, unfamiliar situations.
- You need to have practised many, many exam questions so that you are ready for the type of question that will appear.

The exams test all areas of the syllabus and any weaknesses in your biology will be found out. There must be no holes in your knowledge and understanding. For a Grade A, you must be competent in all areas.

For a Grade C

You must have a reasonable grasp of biology but you may have weaknesses in several areas and you will be unsure of some of the reasons for the biology.

- Many Grade C candidates are just as good at answering questions as the Grade A students but holes and weaknesses often show up in just some topics.
- To improve, you will need to master your weaknesses and you must prepare thoroughly for the exam. You must become a better all-rounder.

For a Grade E

You cannot afford to miss the easy marks. Even if you find biology difficult to understand and would be happy with a Grade E, there are plenty of questions in which you can gain marks.

- You must memorise all definitions.
- You must practise exam questions to give yourself confidence that you do know some biology. In exams, answer the parts of questions that you know first. You must not waste time on the difficult parts. You can always go back to these later.
- The areas of biology that you find most difficult are going to be hard to score on in exams. Even in the difficult questions, there are still marks to be gained. Show your working in calculations because credit is given for a sound method. You can always gain some marks if you get part of the way towards the solution.

What marks do you need?

The table below shows how your average mark is transferred into a grade.

average	80%	70%	60%	50%	40%
grade	A	B	C	D	E

Four steps to successful revision

Step 1: Understand

- Study the topic to be learned slowly. Make sure you understand the logic or important concepts.
- Mark up the text if necessary – underline, highlight and make notes.
- Re-read each paragraph slowly.

Step 2: Summarise

- Now make your own revision note summary:
  What is the main idea, theme or concept to be learnt?
  What are the main points? How does the logic develop?
  Ask questions: Why? How? What next?
- Use bullet points, mind maps, patterned notes.
- Link ideas with mnemonics, mind maps, crazy stories.
- Note the title and date of the revision notes
  (e.g. Biology: Cells, 3rd March).
- Organise your notes carefully and keep them in a file.

This is now in short-term memory. You will forget 80% of it if you do not go to Step 3.
GO TO STEP 3, but first take a 10 minute break.

Step 3: Memorise

- Take 25 minute learning 'bites' with 5 minute breaks.
- After each 5 minute break test yourself:
  Cover the original revision note summary.
  Write down the main points.
  Speak out loud (record on tape).
  Tell someone else.
  Repeat many times.

The material is well on its way to long-term memory.
You will forget 40% if you do not do step 4. GO TO STEP 4

Step 4: Track/Review

- Create a Revision Diary (one A4 page per day).
- Make a revision plan for the topic, e.g. 1 day later, 1 week later, 1 month later.
- Record your revision in your Revision Diary, e.g.
  Biology: Cells, 3rd March 25 minutes
  Biology: Cells, 5th March 15 minutes
  Biology: Cells, 3rd April 15 minutes
  ... and then at monthly intervals.

Chapter 1

Biological molecules

The following topics are covered in this chapter:

- *Essential elements*
- *Carbohydrates*
- *Lipids*
- *Proteins*
- *The importance of water to life*
- *Biochemical tests and chromatography*

1.1 Essential elements

After studying this section you should be able to:

- *describe the range of elements essential for life*
- *recall the four major elements and how they are linked to form biological molecules*

LEARNING SUMMARY

Elements required for living processes

CCEA 1.1
WJEC 1.1

Living things are based on a total of 16 elements out of the 92 which exist on Earth. Over 99% of the biomass of organisms is composed of just 4 key elements: **carbon**, **hydrogen**, **oxygen**, and **nitrogen**.

Carbon is the most important element because of its following properties:

- carbon atoms bond with each other in long chains
- the chains can be branched or even joined up as rings
- the carbon atoms bond with other important elements such as hydrogen, oxygen, nitrogen, sulfur, calcium and phosphorus.

The linking of carbon to carbon in long chains forms the backbone of important structural molecules. Electrons not used in the bonding of carbon to carbon are shared with other elements, such as hydrogen, oxygen and nitrogen.

Element	*Percentage (approximate)*
Carbon	9.5
Hydrogen	63.0
Oxygen	25.5
Nitrogen	1.4
Calcium	0.32
Potassium	0.06
Phosphorus	0.20
Chlorine	0.03
Sulfur	0.05
Sodium	0.03

Trace elements

The table in the margin shows a range of elements found in the human body.

The ten elements in the table are needed in relatively large amounts in living organisms. They are examples of **major elements**. Organisms also need other elements in smaller amounts. These are called **trace elements**. They include cobalt, copper and zinc. These elements are essential, but in higher concentrations, they may be poisonous.

Many of the elements are very reactive and are present in the body as mineral ions. Examples include Mg^{2+}, Fe^{2+}, Ca^{2+} and PO^{4+}.

1.2 Carbohydrates

LEARNING SUMMARY

After studying this section you should be able to:

- *recall the main elements found in carbohydrates*
- *recall the structure of glucose, fructose, lactose, sucrose, starch, glycogen and cellulose*
- *recall the role of glucose, starch, glycogen, cellulose and pectin*

Structure of carbohydrates

AQA	1.2 2.4
CCEA	1.1
EDEXCEL	1.3-4
OCR	2.1.1
WJEC	1.1

Monosaccharides

All carbohydrates are formed from the elements carbon (C), hydrogen (H) and oxygen (O). The formula of a carbohydrate is always $(CH_2O)_n$. The n represents the number of times the basic CH_2O unit is repeated, e.g. where n = 6 the molecular formula is $C_6H_{12}O_6$. This is the formula shared by glucose and other simple sugars like fructose. These simple sugars are made up from a single sugar unit and are known as **monosaccharides**.

The molecular formula, $C_6H_{12}O_6$, does not indicate how the atoms bond together. Bonded to the carbon atoms are a number of –H and –OH groups. Different positions of these groups on the carbon chain are responsible for different properties of the molecules. The structural formulae of α and β glucose are shown below.

These molecules are mirror images of each other. When molecules have the same molecular formula but different structural formulae, they are known as **isomers**. Isomers have different properties to each other.

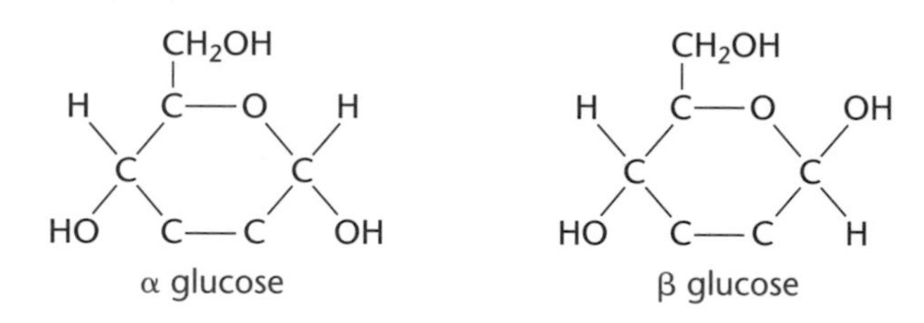

Glucose is so small that it can pass through the villi and capillaries into our bloodstream. The molecules subsequently release energy as a result of respiration. Simple glucose molecules are capable of so much more than just releasing energy. They can combine with others to form bigger molecules.

Disaccharides

Each glucose unit is known as a **monomer** and is capable of linking others. This diagram shows two molecules of α glucose forming a **disaccharide**.

In your examinations look for different monosaccharides being given, like fructose or β glucose. You may be asked to show how they bond together. The principle will be exactly the same.

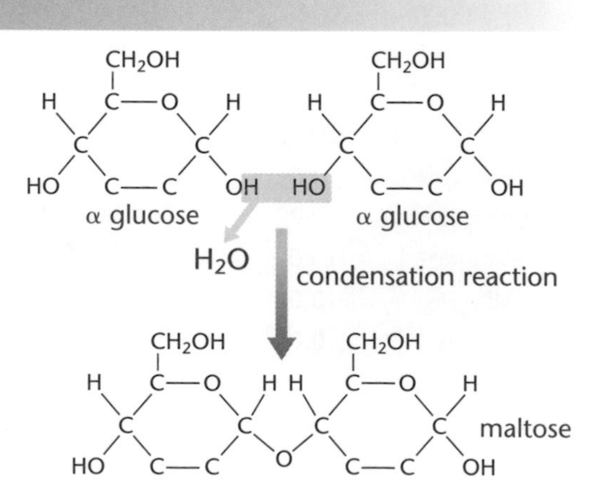

A **condensation** reaction means that as two carbohydrate molecules bond together a water molecule is produced. The link formed between the two glucose molecules is known as a **glycosidic bond**.

A glycosidic bond can also be broken down to release separate monomer units. This is the opposite of the reaction shown above. Instead of water being given off, a water molecule is needed to break each glycosidic bond. This is called **hydrolysis** because water is needed to split up the bigger molecule.

'Lysis' literally means **'splitting'**. In hydrolysis water is needed in the reaction to break down the molecule.

+ H_2O → Hydrolysis → β galactose + α glucose

Different disaccharides are made by joining together different monosaccharides.

Disaccharide	*Component monosaccharides*
lactose	glucose + galactose
maltose	glucose + glucose
sucrose	glucose + fructose

Polysaccharides

Like disaccharides, **polysaccharides** consist of monomer units linked by the glycosidic bond. However, instead of just two monomer units they can have many. Chains of these 'sugar' units are known as **polymers**. These larger molecules have important structural and storage roles.

Starch is a polymer of the sugar, α glucose. The diagram below shows part of a starch molecule.

Notice the five glycosidic bonds on just a small part of a starch molecule.

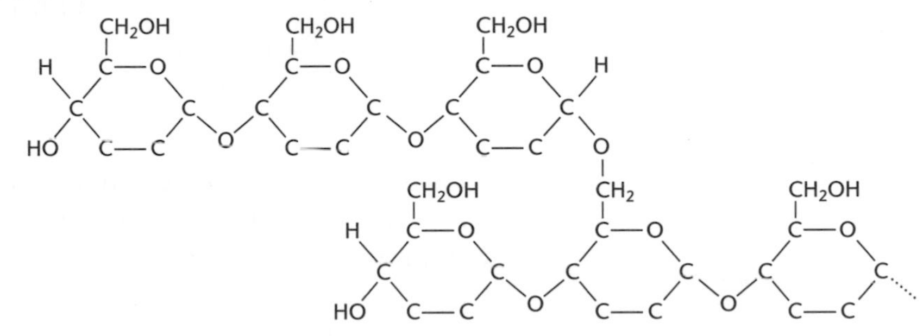

part of a branched section of a starch molecule

This type of starch molecule is called **amylopectin** and it has a branched structure. Starch also contains **amylose**. This does not contain branches but the chain of glucose units forms a helix. **Glycogen** is similar in structure to amylopectin but with more branches. **Cellulose** is also a polymer of glucose units, but this time the units are β glucose.

How useful are polysaccharides?

- **Starch** is stored in organisms as a future energy source, e.g. potato has a high starch content to supply energy for the buds to grow at a later stage.
- **Glycogen** is stored in the liver, which releases glucose for energy in times of low blood sugar.

Both starch and glycogen are insoluble which enables them to remain inside cells.

The many branches in the amylopectin molecule means that enzymes can digest the molecule rapidly.

- **Cellulose** has long molecules which help form a tough protective layer around plant cells, the cell wall. Each cellulose molecule has up to 10 000 β glucose units. Each molecule can form cross-links with other cellulose molecules forming fibres. This makes cellulose fibres very strong.
- **Pectins** are used alongside cellulose in the cell wall. They are polysaccharides which are bound together by calcium pectate.

Pectins help cells to bind together.

Together the cellulose and pectins give exceptional mechanical strength. The cell wall is also permeable to a wide range of substances.

1.3 Lipids

After studying this section you should be able to:

- *recall the main elements found in lipids*
- *recall the structure of lipids*
- *distinguish between saturated and unsaturated fats*
- *recall the role of lipids*

LEARNING SUMMARY

What are lipids?

AQA	1.3
CCEA	1.1
EDEXCEL	1.5
OCR	2.1.1
WJEC	1.1

Lipids include **oils, fats and waxes.** They consist of exactly the same elements as carbohydrates, i.e. carbon (C), hydrogen (H) and oxygen (O), but their proportion is different. There is always a high proportion of carbon and hydrogen, with a small proportion of oxygen.

The diagram below shows the structural formula of the most common type of lipid called a **triglyceride.**

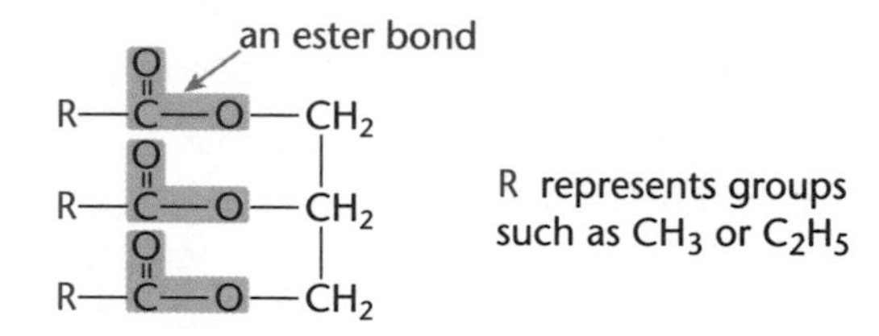

a **triglyceride** fat

Triglycerides are formed when fatty acids react with glycerol. During this reaction water is produced, a further example of a condensation reaction. The essential bond is the **ester bond.**

Note that water is produced during triglyceride formation. This is another example of a condensation reaction. Different triglyceride fats are formed from different fatty acids.

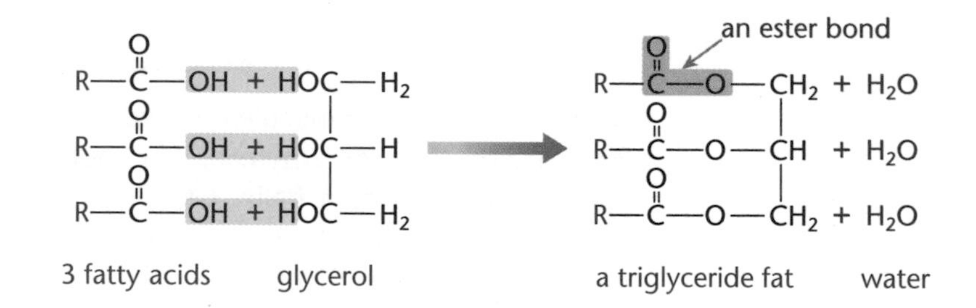

Triglycerides can be changed back into the original fatty acids and glycerol. Enzymes are needed for this transformation together with water molecules. Remember, an enzyme reaction which requires water to break up a molecule is known as **hydrolysis.**

What are saturated and unsaturated fats?

The answer lies in the types of fatty acid used to produce them.

The hydrocarbon chains are so long that they are often represented by the acid group (–COOH) and a zig-zag line.

unsaturated

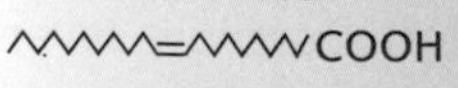

∧∧∧∧=∧∧∧∧COOH

saturated

∧∧∧∧∧∧∧∧∧COOH

stearic acid

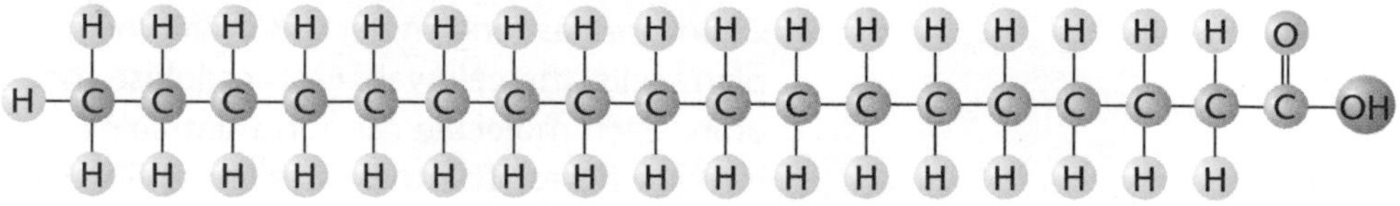

a **saturated** fatty acid

Saturated fats lead to cholesterol build up in the blood vessels so many people try to avoid them in their diet. Blocked vessels cause heart attacks.

Note that when saturated fatty acids react with glycerol they make saturated fats. Conversely, unsaturated fatty acids make unsaturated fats.

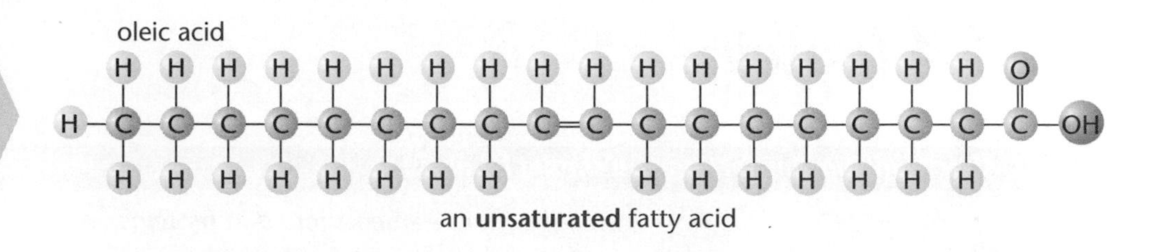

an **unsaturated** fatty acid

> Saturated fatty acids have no C=C (double bonds) in their hydrocarbon chain, but unsaturated fatty acids do. This is the difference.
>
> KEY POINT

How useful are lipids?

Like carbohydrates, lipids are used as an **energy** supply, but a given amount of lipid releases more energy than the same amount of carbohydrate. Due to their **insolubility** in water and **compact** structure, lipids have long-term **storage** qualities. Adipose cells beneath our skin contain large quantities of fat which **insulate** us and help to maintain body temperature. Fat gives **mechanical** support around our soft organs and even gives **electrical insulation** around our nerve axons.

Never give an ambiguous answer like this: 'fats keep an animal warm'. The examiner does not know whether you are referring to energy release by mitochondria or the insulation property. Be specific, otherwise you will not be credited.

An aquatic organism such as a dolphin has a large fat layer which:

- is an energy store
- is a thermal insulator
- helps the animal remain buoyant.

The most important role of lipids is their function in cell membranes. To fulfil these functions a triglyceride fat is first converted into a **phospholipid**.

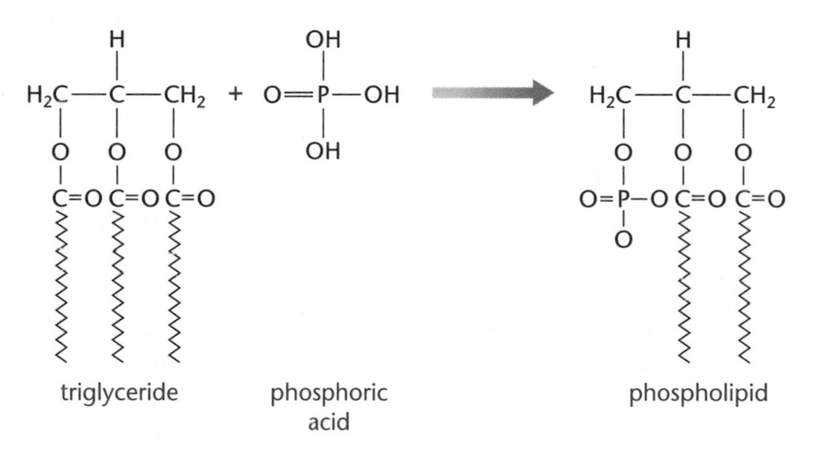

Phosphoric acid replaces one of the fatty acids of the triglyceride. The new molecule, the phospholipid, is a major component of cell membranes. Cell membranes also contain the lipid cholesterol. The diagram below represents a phospholipid.

Remember this simplified diagram when you learn about the cell membrane.

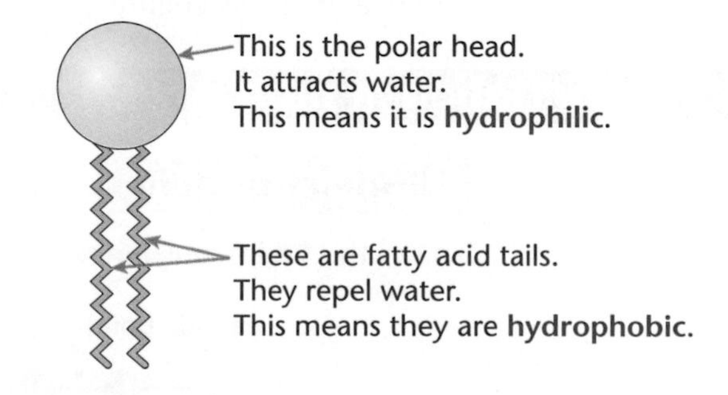

a phospholipid

1.4 Proteins

After studying this section you should be able to:

- recall the main elements found in proteins
- recall how proteins are constructed
- recall the structure of proteins
- recall the major functions of proteins

LEARNING SUMMARY

The building blocks of proteins

AQA	1.2
CCEA	1.1
EDEXCEL	2.7
OCR	2.1.1
WJEC	1.1

Like carbohydrates and lipids, proteins contain the elements carbon (C), hydrogen (H) and oxygen (O), but in addition they **always** contain **nitrogen** (N). Sulfur is also often present.

Before understanding how proteins are constructed, the structure of **amino acids** should be noted. The diagram below shows the general structure of an amino acid.

Just like the earlier carbohydrate and lipid molecules, '**R**' represents groups such as $-CH_3$ and $-C_2H_5$. There are about 20 commonly found amino acids but you will not need to know them all. Instead, learn the basic structure shown opposite.

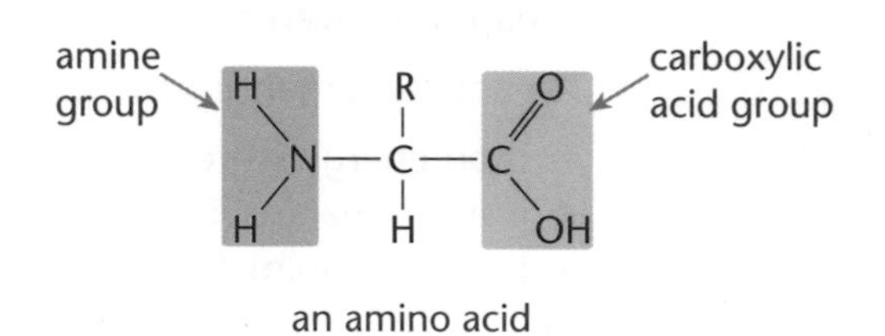

an amino acid

How is a protein constructed?

The process begins by amino acids bonding together. The diagram shows two amino acids being joined together by a **peptide bond**.

This is another example of a condensation reaction as water is produced as the dipeptide molecule is assembled.

Note that the peptide bonds can be broken down by a hydrolysis reaction.

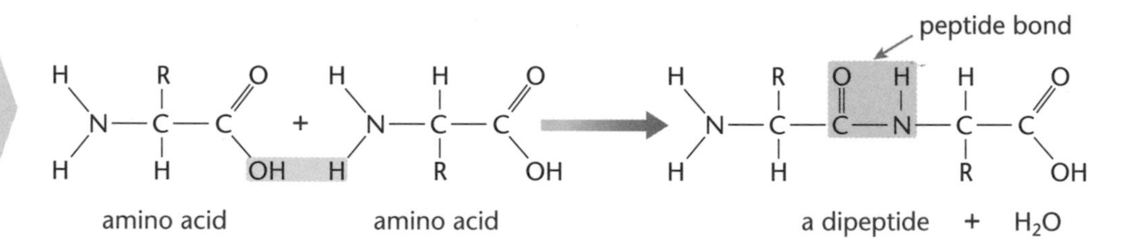

When many amino acids join together a long-chain **polypeptide** is produced. The linking of amino acids in this way takes place during protein synthesis (see page 83). There are around 20 different amino acids. Organisms join amino acids in different linear sequences to form a variety of polypeptides, then build these polypeptides into complex molecules, the **proteins**. Humans need eight essential amino acids as adults and ten as children, all the others can be made inside the cells.

The sequence of amino acids along a polypeptide is controlled by another complex molecule, DNA (see the genetic code, page 82).

Levels of organisation in proteins

AQA	1.2
CCEA	1.1
EDEXCEL	2.7
OCR	2.1.1
WJEC	1.1

Primary protein structure

This is the **linear sequence** of amino acids.

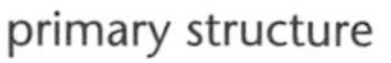

primary structure

Secondary protein structure

Polypeptides become twisted or coiled. These shapes are known as the **secondary structure**. There are two common secondary structures; the α-helix and the β-pleated sheet.

Both secondary structures give additional strength to proteins. The α-helix helps make tough fibres like the protein in your nails, e.g. keratin. The β-pleated sheet helps make the strength-giving protein in silk, fibroin. Many proteins are made from both α-helix and β-pleated sheet.

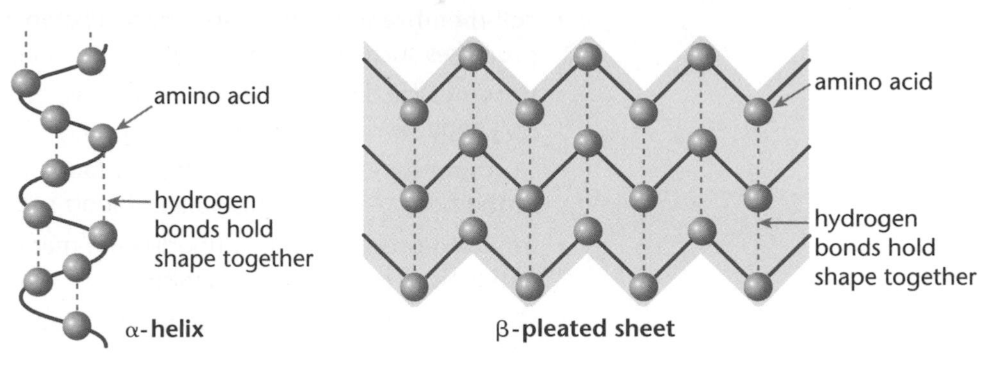

The polypeptides are held in position by hydrogen bonds. In both α-helices and β-pleated sheets the **C = O** of one amino acid bonds to the **H–N** of an adjacent amino acid like this: C = O --- H–N.

----- = hydrogen bonds

An α-helix is a tight, twisted strand; a β-pleated sheet is where a zig-zag line of amino acids bonds with the next, and so on. This forms a sheet or ribbon shape.

The protein shown only achieves a secondary structure as the simple α-helix polypeptides do not undergo further folding.

This is the structure of a fibrous protein. It is made of three α-helix polypeptides twisted together.

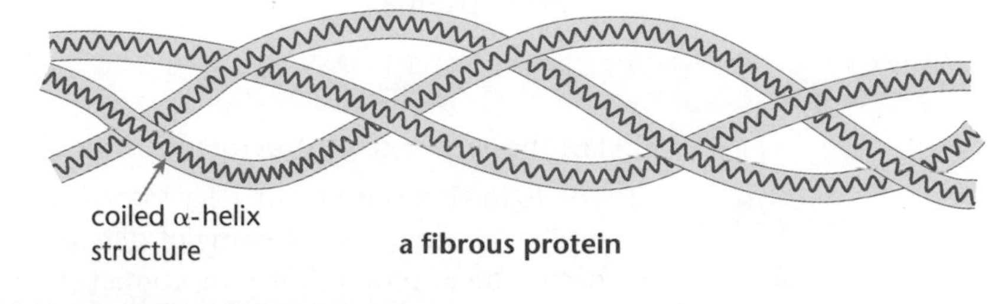

a fibrous protein

Tertiary protein structure

This is when a polypeptide is **folded** into a **precise** shape. The polypeptide is held in 'bends' and 'tucks' in a **permanent** shape by a range of bonds including:

- **disulfide** bridges (sulfur–sulfur bonds)
- hydrogen bonds
- ionic bonds
- hydrophobic and hydrophilic interactions.

Some proteins are folded up into a spherical shape. They are called **globular proteins**. They are soluble in water. Other proteins, called **fibrous proteins**, form long chains. They are insoluble and usually perform structural functions.

Note that the specific contours of proteins have extremely significant roles in life processes. (See enzymes page 46.)

Quaternary protein structure

Some proteins consist of **different polypeptides** bonded together to form extremely intricate shapes. A haemoglobin molecule is formed from four separate polypeptide chains. It also has a haem group, which contains iron. This inorganic group is known as a **prosthetic group** and in this instance aids oxygen transport.

This is the structure of a **globular** protein. It is made of an α-helix and a β-pleated sheet. Precise shapes are formed with specific contours.

Note that some proteins do not have a quaternary structure. If they consist of just one folded polypeptide then they are classified as having tertiary structure. If they are simple fibres of α-helices or β-pleated sheets then they have only secondary protein structure.

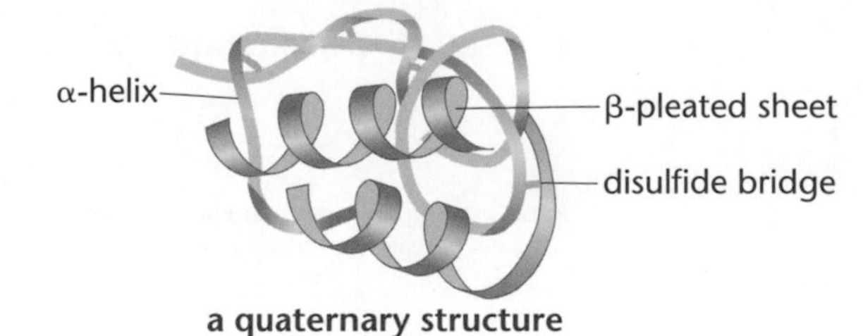

a quaternary structure

How useful are proteins?

AQA	1.2
CCEA	1.1
EDEXCEL	2.7
OCR	2.1.1
WJEC	1.1

Proteins can be just as beneficial as carbohydrates and lipids in releasing energy. Broken down into their component amino acids, they liberate energy during respiration. The list below shows **important** uses of proteins:

- **cell-membrane proteins** transport substances across the membrane for processes such as facilitated diffusion and active transport
- **enzymes** catalyse biochemical reactions, e.g. pepsin breaks down protein into polypeptides
- **hormones** are passed through the blood and trigger reactions in other parts of the body, e.g. insulin regulates blood sugar
- **immuno-proteins**, e.g. antibodies are made by lymphocytes and act against antigenic sites on micro-organisms
- **structural proteins** give strength to organs, e.g. collagen makes tendons tough
- **transport proteins**, e.g. haemoglobin transports oxygen in the blood
- **contractile proteins**, e.g. actin and myosin help muscles shorten during contraction
- **storage proteins**, e.g. aleurone in seeds helps germination, and casein in milk helps supply valuable protein to babies
- **buffer proteins**, e.g. blood proteins, due to their charge, help maintain the pH of plasma.

Progress check

1 List the sequence of structures in a globular protein such as haemoglobin.

2 The following statements refer to proteins used for different functions in the body. The list gives the name of different types of protein. Match the name of each type of protein with the correct statement.

(i) transport proteins
(ii) immuno-proteins
(iii) storage proteins
(iv) buffer proteins
(v) cell-membrane proteins
(vi) contractile proteins
(vii) enzymes
(viii) structural proteins
(ix) hormones

A haemoglobin is used to transport oxygen in blood.
B aleurone in seeds is a source of amino acids as it is broken down during germination.
C actin and myosin help muscles shorten during contraction.
D antibodies made by lymphocytes against antigens.
E blood proteins, due to their charge, help maintain the pH of plasma.
F used to transport substances across the membrane for processes such as facilitated diffusion.
G passed through blood, used to trigger reactions in other parts of the body, e.g. FSH stimulates a primary follicle.
H used to catalyse biochemical reactions, e.g. amylase breaks down starch into maltose.
I used to give strength to organs, e.g. collagen makes tendons tough.

1 primary structure: amino acids linked in a linear sequence; secondary structure: α-helix or β-pleated sheet; tertiary structure: further folding of polypeptide held by disulfide bridges, ionic bonds, and hydrogen bonds; quaternary structure: two or more polypeptides bonded together.
2 A (i), B (iii), C (vi), D (ii), E (iv), F (v), G (ix), H (vii), I (viii).

1.5 The importance of water to life

After studying this section you should be able to:

- *recall the properties of water*
- *recall the functions of water*

LEARNING SUMMARY

Properties and uses of water

CCEA	1.1.1
EDEXCEL	1.2
OCR	2.1.1
WJEC	1.1

Try to learn all of the functions of water molecules given in the list. Water is used in so many ways that the chance of being questioned on the topic is high.

Water is essential to living organisms. The list below shows some of its properties and uses.

- **Hydrogen bonds** are formed between the oxygen of one water molecule and the hydrogen of another. As a result of this water molecules have an attraction for each other known as **cohesion**.
- **Cohesion** is responsible for surface tension which enables aquatic insects like pond skaters to walk on a pond surface. It also aids capillarity, the way in which water moves through xylem in plants.
- Water is a **dipolar** molecule, which means that the oxygen has a slight negative charge at one end of the molecule, and each hydrogen has a slight positive charge at the other end.
- Other **polar** molecules dissolve in water. The different charges on these molecules enable them to fit into water's hydrogen bond structure. Ions in solution can be transported or can take part in reactions. Polar substances can dissolve in water and are called **hydrophilic**. Non-polar substances cannot dissolve in water and are **hydrophobic**.
- Water is used in **photosynthesis**, so it is necessary for the production of glucose. This in turn is used in the synthesis of many chemicals.
- Water helps in the **temperature regulation** of many organisms. It enables the cooling down of some organisms. Owing to a **high latent heat of vaporisation**, large amounts of body heat are needed to evaporate a small quantity of water. Organisms like humans cool down effectively but lose only a small amount of water in doing so.
- A relatively high level of heat is needed to raise the temperature of water by a small amount due to its **high specific heat capacity**. This enables organisms to control their body temperature more effectively.
- Water is a solvent for ionic compounds. A number of the essential elements required by organisms are obtained in ionic form, e.g.:
  (a) plants absorb nitrate ions (NO_3^-) and phosphate ions (PO_4^-) in solution
  (b) animals intake sodium ions (Na^+) and chloride ions (Cl^-).

1.6 Biochemical tests and chromatography

After studying this section you should be able to:

- *describe biochemical tests for carbohydrates, lipids and proteins*
- *describe the separation and identification of molecules by chromatography*

LEARNING SUMMARY

Biochemical tests

AQA	1.2
CCEA	1.1
OCR	2.1.1
WJEC	1.1

Tests for carbohydrates in the laboratory

Benedict's test used to identify reducing sugars (monosaccharides and some disaccharides)

- Add Benedict's solution to the chemical sample and heat.
- The solution changes from blue to brick-red or yellow if a reducing sugar is present.

All the biochemical tests need to be learned. This work is good value because they are regularly tested in 2 or 3 mark question components.

Non-reducing sugar test used to test for non-reducing sugars, e.g. the disaccharide, sucrose

- First a Benedict's test is performed.
- If the Benedict's test is negative, the sample is hydrolysed by heating with hydrochloric acid, then neutralised with sodium hydrogen carbonate.
- This breaks the glycosidic bond of the disaccharide, releasing the monomers.
- A second Benedict's test is performed which will be positive because the monomers are now free.

Starch test

- Add iodine solution to the sample.
- If starch is present the colour changes to blue-black.

Tests for lipids in the laboratory

Emulsion test used to identify fats and oils

- Add ethanol to the sample, shake, then pour the mixture into water.
- If fats or oils are present then a white emulsion appears at the surface.

Tests for proteins in the laboratory

Biuret test used to identify any protein

- Add dilute sodium hydroxide and dilute copper sulfate to the sample.
- A violet colour appears if a protein is present.

Quantitative tests

The above tests can be used to see if the different chemicals are present or absent. They are called **quantitative tests**.

They do not provide any information about how much of the chemical is present unless they are modified. One way to do this is to measure the intensity of the colour produced. This can be done with a machine called a **colorimeter**. It shines a light through the solution and measures how much light gets through. The lighter the colour, the more light passes through and so the less chemical is present. This is a **quantitative test**.

Chromatography

CCEA 1.1

This technique is used to separate out the components in a mixture such as chlorophyll, and can be used to help identify substances. The method is outlined below:

- a spot of the substance is placed on chromatography paper and left to dry
- the paper is suspended in a solvent such as propanone
- as the solvent molecules move through the paper the components begin to move up the paper,
  big molecules move slower than small ones
- the small solvent molecules move through the paper faster than any of the components of the substance
- the substance separates out into different spots or bands.

You need to remember that this technique separates substances in terms of the relative size of the molecules.

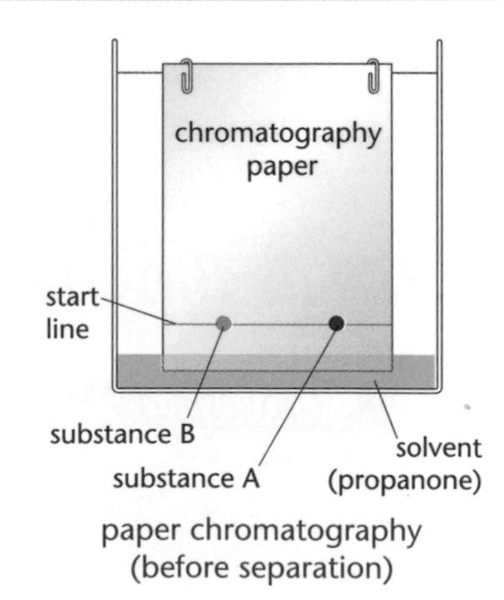

paper chromatography (before separation)

The longer the chromatogram is left after the start, the higher the spots or bands ascend. For this reason every compound has its R_f value calculated. However long the chromatogram is left, the R_f value is the same when using the same solvent.

R_f value of a substance

This is calculated after the distances moved by compounds and solvent up the chromatogram have been measured.

$$R_f \text{ value} = \frac{\text{distance moved by substance}}{\text{distance moved by solvent front}}$$

Different compounds show up as different coloured bands. The technique shows the number of compounds in the mixture. When this method is used on chlorophyll, five colours separate out.

solvent front

distance moved by solvent

substance B had two component compounds

substance A had three component compounds

Paper chromatography (after separation)

Progress check

A chromatogram was prepared for substance X. Five different spots were noted on the chromatogram.

1 What does this indicate?

2 What is the equation used to calculate the R_f value of a spot?

1 Substance X consisted of five different compounds.

2 R_f value = $\frac{\text{distance moved by substance}}{\text{distance moved by solvent front}}$

Sample questions and model answers

Remember that you will be given molecule structures. These stimulate your memory which helps you work out the answer.

1 Below are the structures of two glucose molecules.

(a) Complete the equation to show how the molecules react to form a glycosidic bond and the molecule produced.

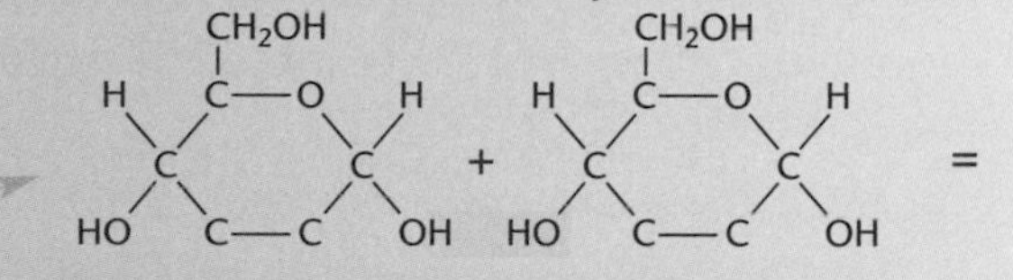

CH₂OH CH₂OH H C—O H H C—O H C C C C HO C—C O C—C OH Glycosidic bond

maltose + H_2O

(b) Which form of glucose molecules is shown? Give a reason for your answer. [2]

α glucose, because the –OH groups on carbon atom 1 are down

The correct answer here is condensation. A regular error in questions like this is to give the wrong reaction, i.e. hydrolysis. Revise carefully then you will make the correct choice.

(c) State the type of reaction which takes place when the two molecules shown above react together. [2]

Condensation.

Look out for similar structures in your examinations. The proteins given may be different, but the principles remain the same.

2 The diagram below shows a globular protein consisting of four polypeptide chains.

α-helix

β-pleated sheet

disulfide bridge

(a) Use your own knowledge and the information given to explain how this protein shows primary, secondary, tertiary and quaternary structure. [5]

Primary structure: it is formed from chains of amino acids; it has polypeptides made of a linear sequence of amino acids.
Secondary structure: it has an α-helix, it has a β-pleated sheet.
Tertiary structure: the polypeptides are folded, the folds are held in position by disulfide bridges.
Quaternary structure: there are four polypeptides in this protein. Two or more are bonded together to give a quaternary structure.

Most examinations include at least one biochemical test.

(b) Name and describe a test which would show that haemoglobin is a protein. [3]

The Biuret test.
Take a sample of haemoglobin and add water, sodium hydroxide and copper sulfate.
The colour of the mixture shows as violet or mauve if the sample is a protein.

Practice examination questions

Try all of the questions and check your answers with the mark scheme on page 117.

1 The chromatogram below shows a substance which has been separated into its component compounds.

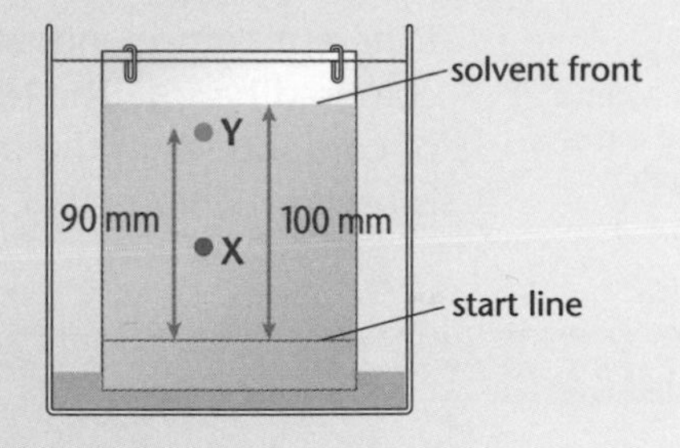

(a) Calculate the R_f value of spot Y. [2]

(b) Which spot contains the biggest molecules? [1]

(c) The chromatogram had been left for six hours after a drop was put on the start line. Why was it important to take the measurement for the calculation of the R_f value of Y before another hour had passed? [1]

2 (a) Complete the equation below to show the breakdown of a triglyceride fat into fatty acids and glycerol. [2]

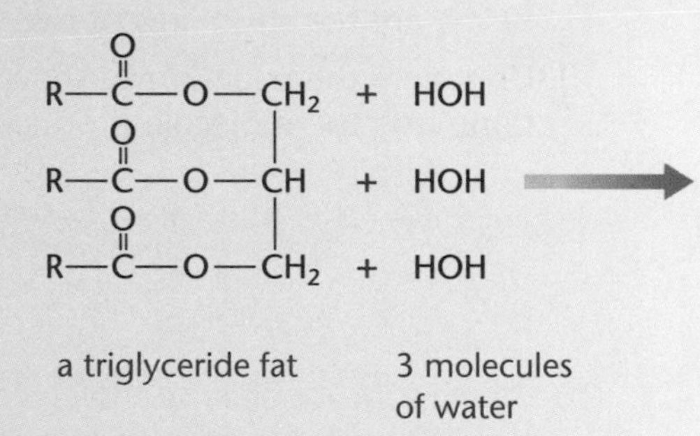

(b) Describe a biochemical test which would show if a sample contained fat. [3]

3 The diagram below shows a polypeptide consisting of 15 amino acids.

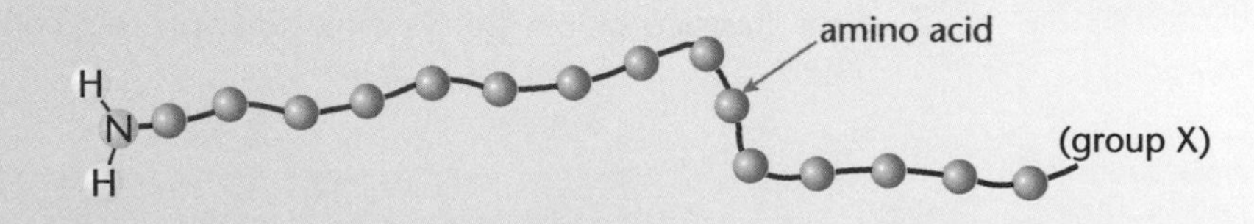

(a) Name the bond between each pair of amino acids in this polypeptide. [1]

(b) What is group X? [1]

(c) Which level of protein structure is shown by this polypeptide? Give a reason for your answer. [2]

4 Explain how the following properties of water are useful to living organisms:

(a) a large latent heat of evaporation [2]

(b) a high specific heat capacity [2]

(c) the cohesive attraction of water molecules for each other. [2]

Chapter 2

Cells

The following topics are covered in this chapter:

- *The ultra-structure of cells*
- *Isolation of cell organelles*
- *Specialisation of cells*

2.1 The ultra-structure of cells

After studying this section you should be able to:

- *identify cell organelles and understand their roles*
- *recall the differences between prokaryotic and eukaryotic cells*

LEARNING SUMMARY

Cell organelles

AQA	1.3 2.4
CCEA	1.5
EDEXCEL	3.2–4 4.2 4.4
OCR	1.1.1
WJEC	1.2

The cell is the basic functioning unit of organisms in which chemical reactions take place. These reactions involve energy release needed to support life and build structures. Organisms consist of one or more cells. The amoeba is composed of one cell, whereas millions of cells make up a human.

KEY POINT

Every cell possesses internal coded instructions to control cell activities and development. Cells also have the ability to continue life by some form of cell division.

Organelles are best seen with the aid of an electron microscope.

The ultra-structure of a cell can be seen using an electron microscope. Sub-cellular units called **organelles** become visible. Each organelle has been researched to help us understand more about the processes of life.

The animal cell and its organelles

The diagram below shows the organelles found in a typical animal cell.

A plant cell has all of the same structures plus:

- a cellulose cell wall
- chloroplasts (some cells)
- a sap vacuole with tonoplast.

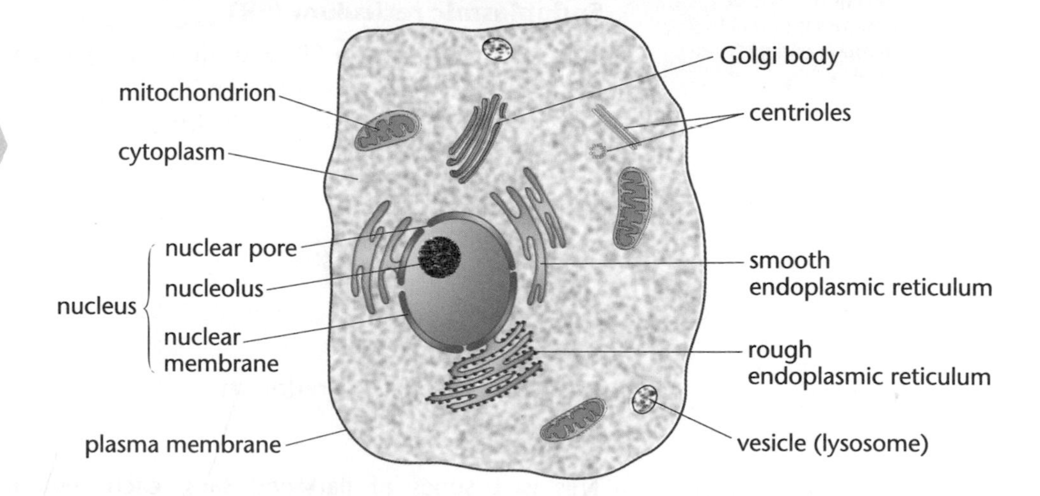

Cell surface (plasma) membrane

This covers the outside of a cell and consists of a **double** layered sheet of phospholipid molecules interspersed with proteins. It separates the cell from the outside environment, gives physical protection and allows the import and export of **selected** chemicals.

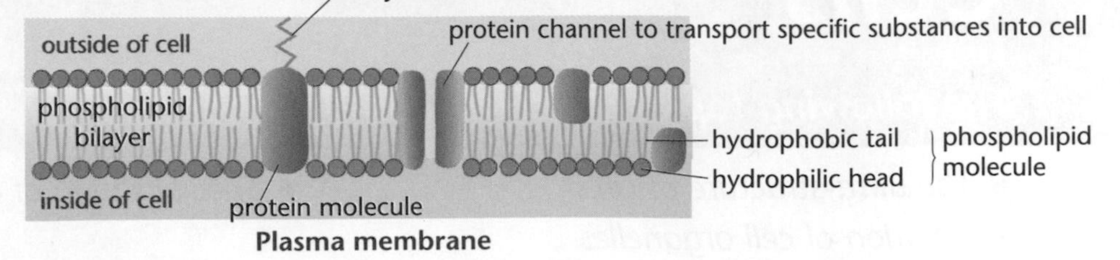

Plasma membrane

Nucleus

This controls all cellular activity using coded instructions located in DNA. These coded instructions enable the cell to make specific proteins. RNA is produced in the nucleus and leaves via the nuclear pores. The nucleus stores, replicates and decodes DNA.

> Be ready to identify all cell organelles in either a diagram or electron micrograph. A mitochondrion is often sausage shaped but the end view is circular. Look out for the internal membranes.

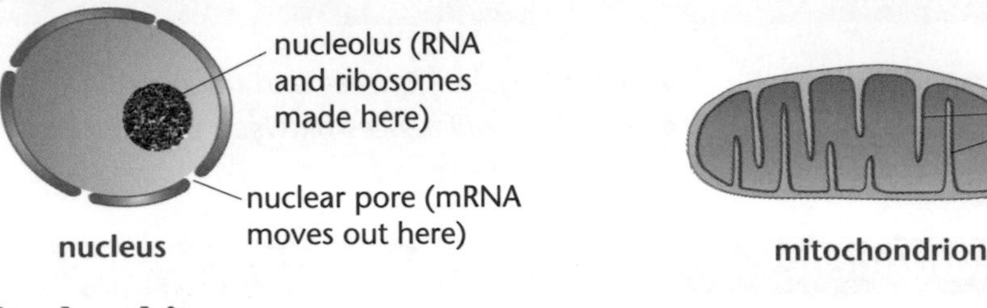

Mitochondria

These consist of a double membrane enclosing a semi-fluid matrix. Throughout the matrix is an internal membrane, folded into cristae. The cristae and matrix contain enzymes which enable this organelle to carry out aerobic respiration. It is the key organelle in the release of energy, making ATP available to the cell.

Mitochondria are needed for many energy requiring processes in the cell, including active transport and the movement of cilia.

> Cytoplasm is often seen as grey and granular. If the image is 'clear' then you are probably looking at a vacuole.

Cytoplasm

Each organelle in a cell is suspended in a semi-liquid medium, the cytoplasm. Many ions are dissolved in it. It is the site of many chemical reactions.

> Look for tiny dots in the cytoplasm. They will almost certainly be ribosomes. A membrane adjacent to a line of ribosomes is probably the rough endoplasmic reticulum.

Ribosomes

There are numerous **ribosomes** in a cell, located along **rough endoplasmic reticulum**. They aid the manufacture of proteins, being the site where mRNA meets tRNA so that amino acids are bonded together.

Endoplasmic reticulum (ER)

This is found as **rough ER** (with ribosomes) and **smooth ER** (without ribosomes). It is a series of folded internal membranes. Substances are transported in the spaces between the ER. The smooth ER aids the synthesis and transport of lipids.

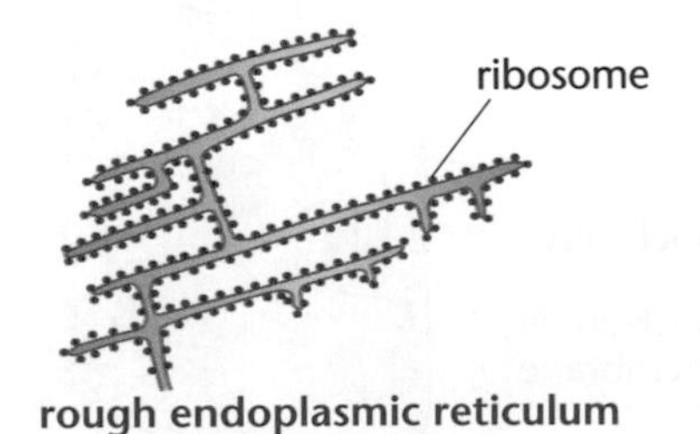

rough endoplasmic reticulum

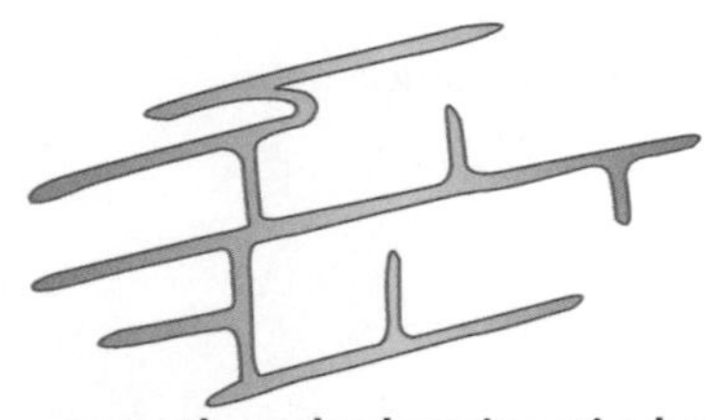

smooth endoplasmic reticulum

> Look for vesicles 'pinching off' the main Golgi sacs.

Golgi body

This is a series of flattened sacs, each separated from the cytoplasm by a membrane. The **Golgi body** is a packaging system where important chemicals become membrane wrapped, forming **vesicles**. The vesicles become detached from the main Golgi sacs, enabling the isolation of chemicals from each other in the cytoplasm. The Golgi body aids the production and secretion of many proteins, carbohydrates and glycoproteins. Vesicle membranes merge with the plasma membrane to enable secretions to take place.

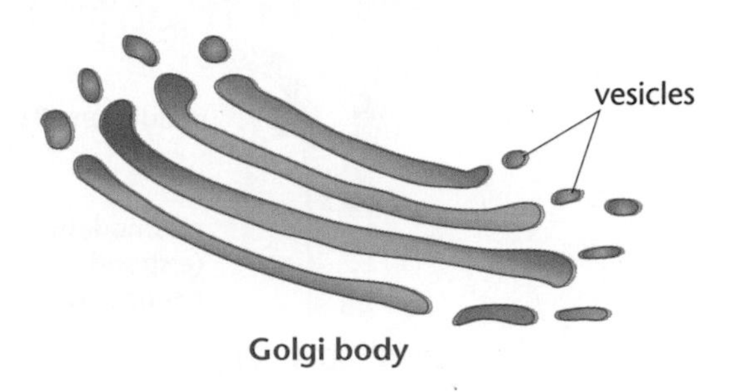

Lysosomes

These are **specialised vesicles** because they contain **digestive enzymes**. The enzymes have the ability to break down proteins and lipids. If the enzymes were free to react in the cytoplasm then cell destruction would result.

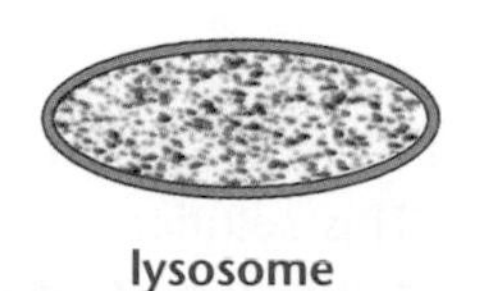
lysosome

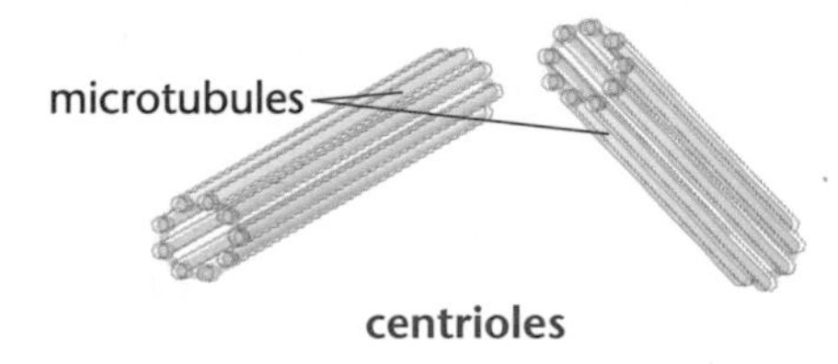

centrioles

Centrioles

In a cell there are two short cylinders which contain **microtubules**. Their function is to aid cell division. During division they move to opposite poles as the spindle develops.

The plant cell and its organelles

All of the structures described for animal cells are also found in plant cells (except the centrioles). Additionally there are three extra structures shown in the diagram below.

> Did you spot the three extra structures in the plant cell? Remember that a root cell under the soil will not possess chloroplasts. Nor does every plant cell above the soil have chloroplasts.

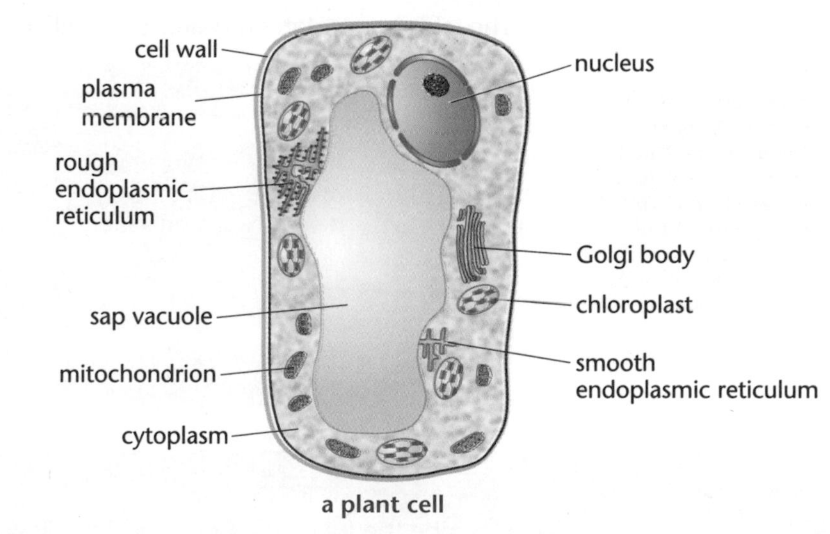

a plant cell

Cell wall

> Cells other than plant cells can have cell walls, e.g. bacteria have polysaccharides other than cellulose.

Around the plasma membrane of plant cells is the cell wall. This is secreted by the cell and consists of cellulose microfibrils embedded in a layer of calcium pectate and hemicelluloses. Between the walls of neighbouring cells calcium pectate cements one cell to the next in multi-cellular plants. Plant cell walls provide a rigid support for the cell but allow many substances to be imported or exported by the cell. The wall allows the cell to build up an effective hydrostatic skeleton. Some plant cells have a cytoplasmic link which crosses the wall. These links of cytoplasm are known as **plasmodesmata**.

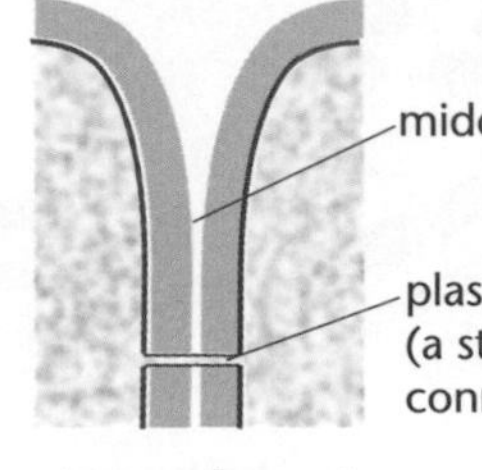

plasmodesmata

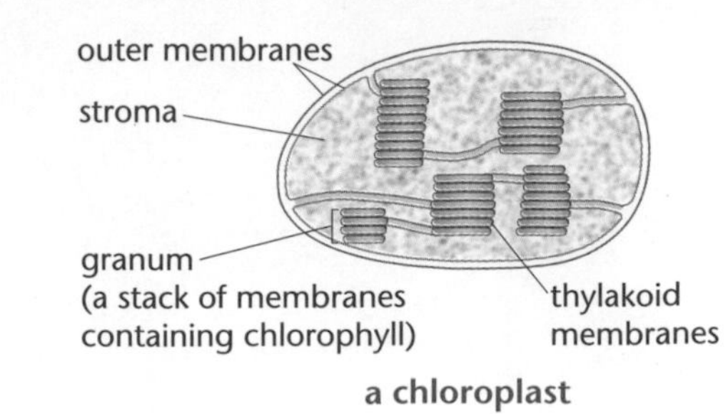

a chloroplast

Chloroplasts

These enable the plant to photosynthesise, making glucose. Each consists of an outer covering of two membranes. Inside are more membranes stacked in piles called **grana**. The membranes enclose a substance vital to photosynthesis, **chlorophyll**. Inside the chloroplast is a matrix known as the **stroma** which is also involved in photosynthesis.

Sap vacuole

This is a large space in a plant cell, containing chemicals such as glucose and mineral ions in water. This solution is the sap. It is surrounded by a membrane known as the **tonoplast**. It is important that a plant cell contains enough water to maintain internal hydrostatic pressure. When this is achieved the cell is turgid, having maximum hydrostatic strength.

Prokaryotic and eukaryotic cells

AQA 1.3
EDEXCEL 3.2
WJEC 1.2
CCEA 1.5

Organisms can be classified into two groups, **prokaryotic** or **eukaryotic** according to their cellular structure.

> **KEY POINT**
> Prokaryotic cells are characteristic of two groups of organisms, bacteria and blue-green algae. Prokaryotic cells are less complex than eukaryotic ones and are considered to have evolved earlier.

The table below states similarities and differences between the two types of organism.

In an examination you will often be given a diagram of a cell from an organism you have not seen before. This is not a problem! The examiners are testing your recognition of the organelles found in typical prokaryotic and eukaryotic organisms.

	Prokaryotic cells	*Eukaryotic cells*
Kingdom	Prokaryotae	Protoctista, Fungi, Animalia, Plantae
Organelles	1 small ribosomes	large ribosomes
	2 DNA present but there is no nuclear membrane	DNA is enclosed in a membrane i.e. has nucleus, mitochondria, (Golgi body vesicles and ER are present)
	3 cell wall present consisting of mucopeptides	cell walls present in plant cells – cellulose cell walls present in fungi – chitin
	4 if cells have flagellae there is no 9+2 microtubule arrangement	if cells have flagellae there is a 9+2 microtubule arrangement

Progress check

1 Describe the function of each of the following cell organelles:

nucleus, centrioles, Golgi body, mitochondria, ribosomes, cell (plasma) membrane

2 Give **three** structural differences between a plant and animal cell.

1 **nucleus** – mRNA is produced in the nucleus with the help of DNA
mitochondria – release energy during aerobic respiration
centrioles – help produce the spindle during cell division
ribosomes – aid the manufacture of proteins, being the site where mRNA meets tRNA so that amino acids are bonded together
Golgi body – is a packaging system where chemicals become membrane wrapped, forming vesicles
Cell (plasma) membrane – gives physical protection to the outside of a cell, allows the import and export of *selected* chemicals.

2 A plant cell has a cellulose cell wall, chloroplasts, and a sap vacuole lined by a tonoplast.

2.2 Isolation of cell organelles

After studying this section you should be able to:

- *understand how cell fractionation and ultracentrifugation are used to isolate cell organelles*
- *understand the principles of light and electron microscopes*
- *understand how microscopic specimens are measured*

LEARNING SUMMARY

Cell fractionation

AQA 1.3

Occasionally it is necessary to isolate organelles to investigate their structure or function, e.g. mitochondria could be used to investigate aerobic respiration away from the cell's internal environment. **Cell fractionation** consists of two processes: **homogenisation** followed by **differential centrifugation**. Cell fractionation depends on the different densities of the organelles.

Technique

Cells are kept:

- cool at around 5°C *(this slows down the inevitable autolysis, destruction by the cell's own enzymes)*
- in an isotonic solution, i.e. equal concentration of substances inside and outside of the cell membrane *(this ensures that the organelles are not damaged by osmosis and can still function)*
- at a specific pH by a buffer solution *(this ensures that the organelles can still function, as they are kept in suitable conditions)*.

Homogenisation

The cells are homogenised in either a pestle homogeniser or a blender. This breaks the cells up releasing the organelles and cytoplasm. Many organelles are not damaged at all by this process. At this stage there is a suspension of mixed organelles.

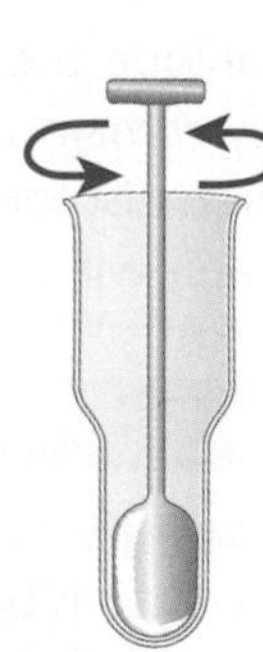

pestle homogeniser

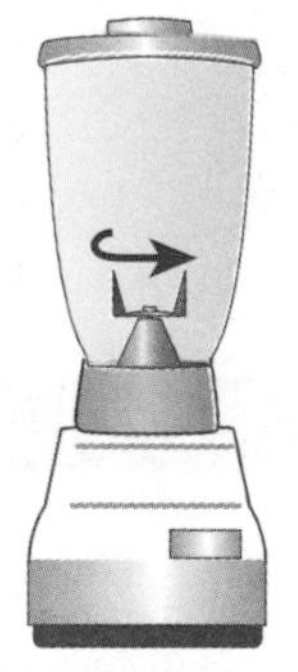

blender

Differential centrifugation

Equal amounts of homogenised tissue are poured into the tubes of an **ultracentrifuge**. This instrument spins the cell contents at a force many times greater than gravity. Organelle separation by this technique is **density dependent**.

600 g means 600 times the force of gravity. If a plant cell was centrifuged at around this speed then chloroplasts would be contained in the sediment as well as nuclei.

First spin

The sample is spun at 600 g for 10 minutes. **Nuclei**, the organelles of greatest density, collect in the **sediment** at the base of the tube. All other cell contents are

The principle of pouring off the supernatant to leave the pure sediment behind can be repeated at the end of each spin. In this way the main organelles can be isolated.

Note that it is much easier to obtain nuclei, because they are isolated in the first spin. Ribosomes are isolated at the final spin.

found in the **supernatant** (the fluid above the sediment). Pouring off the supernatant leaves the sediment of relatively pure nuclei.

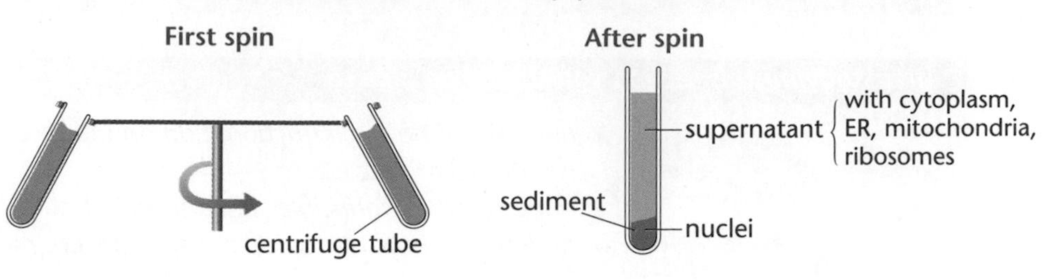

Second spin

The remaining supernatant fluid is spun at 10 000 to 20 000 g for a further 20 minutes. **Mitochondria**, the most dense of the remaining organelles, collect in the sediment. All other cell contents are found in the **supernatant**.

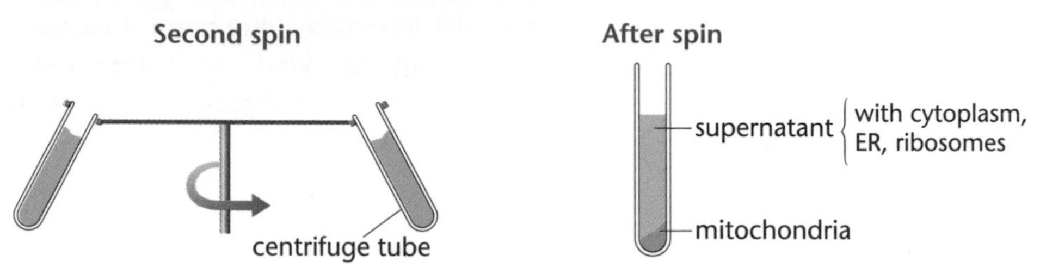

Third spin

The remaining supernatant fluid is spun at 100 000 g for a further 60 minutes. In this sediment are fragments of endoplasmic reticulum and ribosomes. Other cell contents, e.g. cytoplasm and proteins, remain in the supernatant.

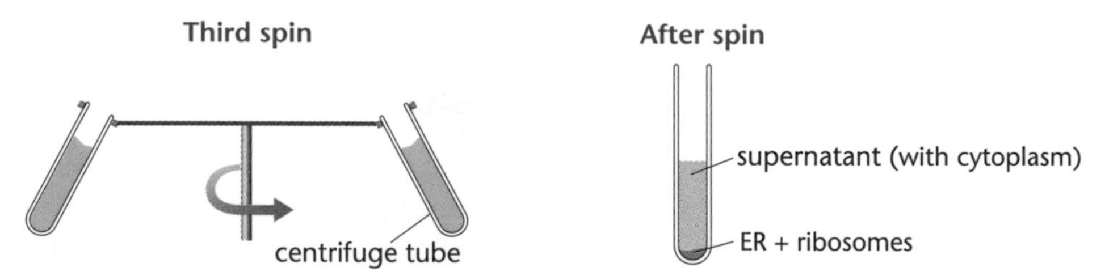

Using this technique the organelles are isolated. By providing them with suitable conditions they remain active for a time and can be used in investigations, e.g. chloroplasts given isotonic solution, suitable (warm) temperature, light, carbon dioxide and water will continue to photosynthesise.

Light and electron microscopy

AQA	1.3
CCEA	1.5
OCR	1.1.1

Microscopes **magnify** the image of a specimen to enable the human eye to see minute objects not visible to the naked eye. **Resolution** of a microscope is the ability to distinguish between two objects as separate entities. At low resolution only one object may be detected. At high resolution two distinct objects are visible. At high resolution the image of such a specimen would show considerable detail.

The light microscope has limited resolution (0.2 μm) due to the wavelength of light so that organelles such as mitochondria, although visible, do not have clarity. Electron microscopes have exceptional resolution. The transmission electron microscope has a high resolution (0.2 to 0.3 nm). This enables even tiny organelles to be seen.

The light microscope

This type of microscope uses white light to illuminate a specimen. The light is focused onto the specimen by a **condensing lens**. The specimen is placed on a microscope slide which is clipped onto a platform, known as the **stage**. The image is viewed via an eyepiece or ocular lens. Overall magnification of the specimen depends on the individual magnification of the eyepiece lens and objective lens. For example, if a specimen is being observed with an eyepiece × 10 and an objective lens of × 40, then the image is 400 times the true size of the specimen.

> **KEY POINT**
> Many specimens need **staining** with chemicals so that tissues and, perhaps, organelles can be seen clearly, e.g. methylene blue is used to stain nuclei. Sometimes more than one stain is used, e.g. in differential staining, so that sub-cellular parts contrast with each other.

To decide which microscope is suitable use the table below.

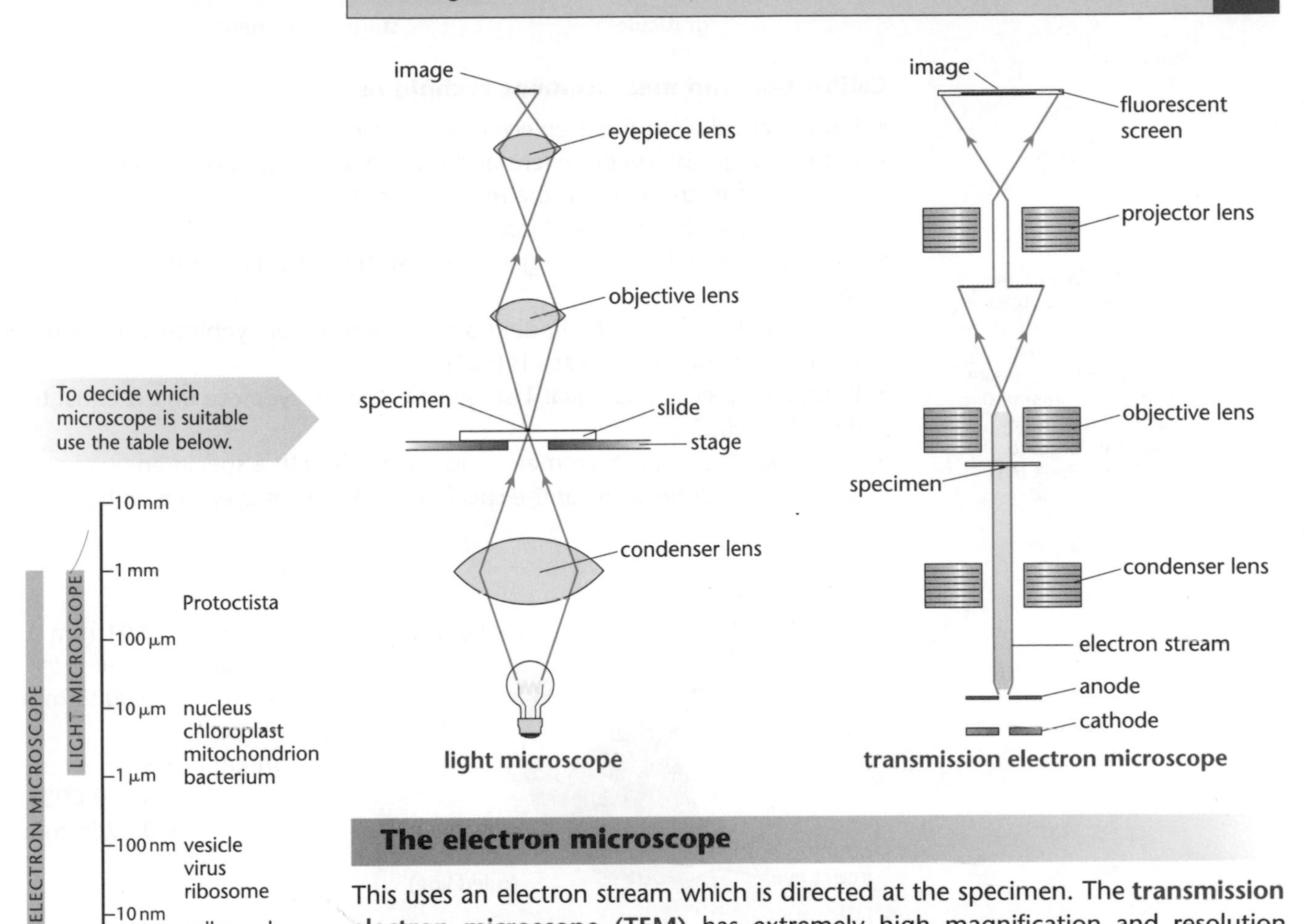

The electron microscope

This uses an electron stream which is directed at the specimen. The **transmission electron microscope (TEM)** has extremely high magnification and resolution properties. Specimens are placed in a vacuum within the microscope, to ensure the electrons do not collide with air molecules and distort the image. **Stains** such as **osmium** and **uranium** salts are used to make organelles distinct. These salts are absorbed by organelles and membranes differentially, e.g. the nuclear membrane absorbs more of the salts than other parts. In this way the nuclear membrane becomes more dense. When the electron beam hits the specimen, electrons are unable to pass through this dense membrane. The membrane shows up as a dark shadow area on the image, because it is in an electron shadow. Cytoplasm allows more electrons to pass through. When these electrons hit the fluorescent screen visible light is emitted.

Artefacts

When microscopic specimens are prepared there are often several chemical and physical procedures. Often the material is dead so changes from the living specimen are expected. Microscopic material should be analysed with care because there may have been some artificial change in the material during preparation, e.g. next to some cells a student might see a series of small circles. They look like eggs but are merely air bubbles. These are **artefacts**; structures alien to the material which should not be interpreted as part of the specimen.

Microscopic measurement

It is sometimes necessary to measure microscopic structures. There are two instruments needed for this process, a **graticule** and **stage micrometer**.

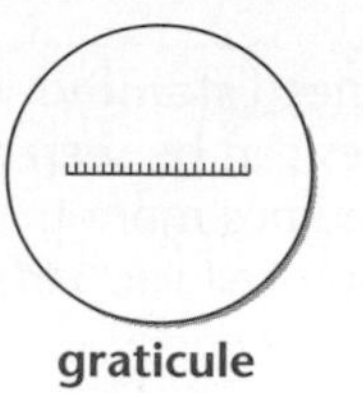
graticule

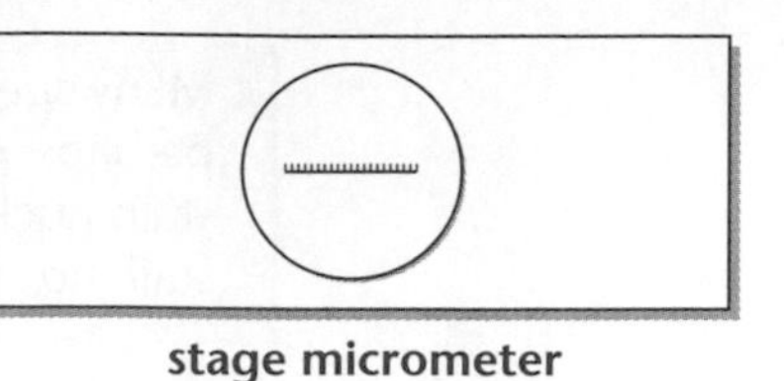
stage micrometer

Calibration and measurement technique

- Put a graticule into the eyepiece of a microscope.
- Look through the eyepiece lens and the graticule line can be seen.
- Put a stage micrometer on the microscope stage.
- Look through the eyepiece lens.
- Line up the ruled line of the graticule with the ruled line of the stage micrometer.
- Calibrate the eyepiece by finding out the number of eyepiece units (e.u.) equal to one stage unit (s.u.); each is 0.01 mm.
- If three eyepiece units equal 1 stage unit, then 1 eyepiece unit is equal to 0.01/3 (0.0033 mm).
- Take away the stage micrometer and replace it with a specimen.
- Measure the dimensions of the specimen in terms of eyepiece units.

Important! Every time you change the objective lens, e.g. move from low power to high power, recalibration is necessary. In an examination this may be tested. Many candidates forget to recalibrate. Don't miss the mark!

Example

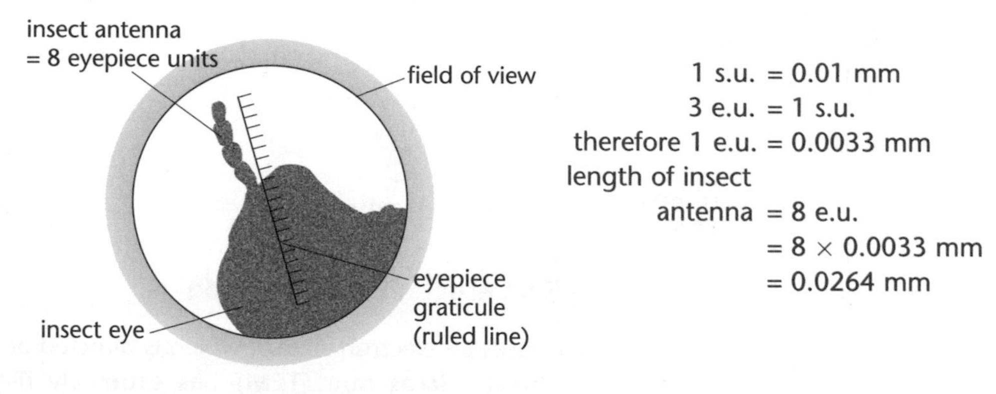

Progress check

1. Describe how the width of an insect's leg could be measured using a microscope, graticule and stage micrometer.
2. A student measured the insect leg but considered that a higher magnification was needed to improve accuracy. What is the significance of changing magnification to the measuring technique?

1 Put a graticule into the eyepiece of a microscope; put a stage micrometer on the microscope stage; look through the eyepiece lens; line up the ruled line of the graticule with the ruled line of the stage micrometer; calibrate the eyepiece by finding out the number of eyepiece units equal to one stage unit; a stage unit is a known length, so an eyepiece unit length can be calculated; take away the stage micrometer and replace with specimen; measure width of insect leg in eyepiece units.
2 Recalibration is needed for each new magnification.

2.3 Specialisation of cells

After studying this section you should be able to:

LEARNING SUMMARY

- *understand cell specialisation and how cells aggregate into tissues and organs*
- *recall a range of cell adaptations*

The earlier parts of this chapter described the structure and function of generalised animal and plant cells. Their features are found in many unicellular organisms where all the life-giving processes are carried out in one cell. Additionally many multicellular organisms exist. A few show no specialisation and consist of repeated identical cells, e.g. Volvox, a colonial alga. Most multicellular organisms exhibit **specialisation**, where different cells are adapted for specific roles.

Cell adaptations

AQA 2.6
OCR 1.1.3

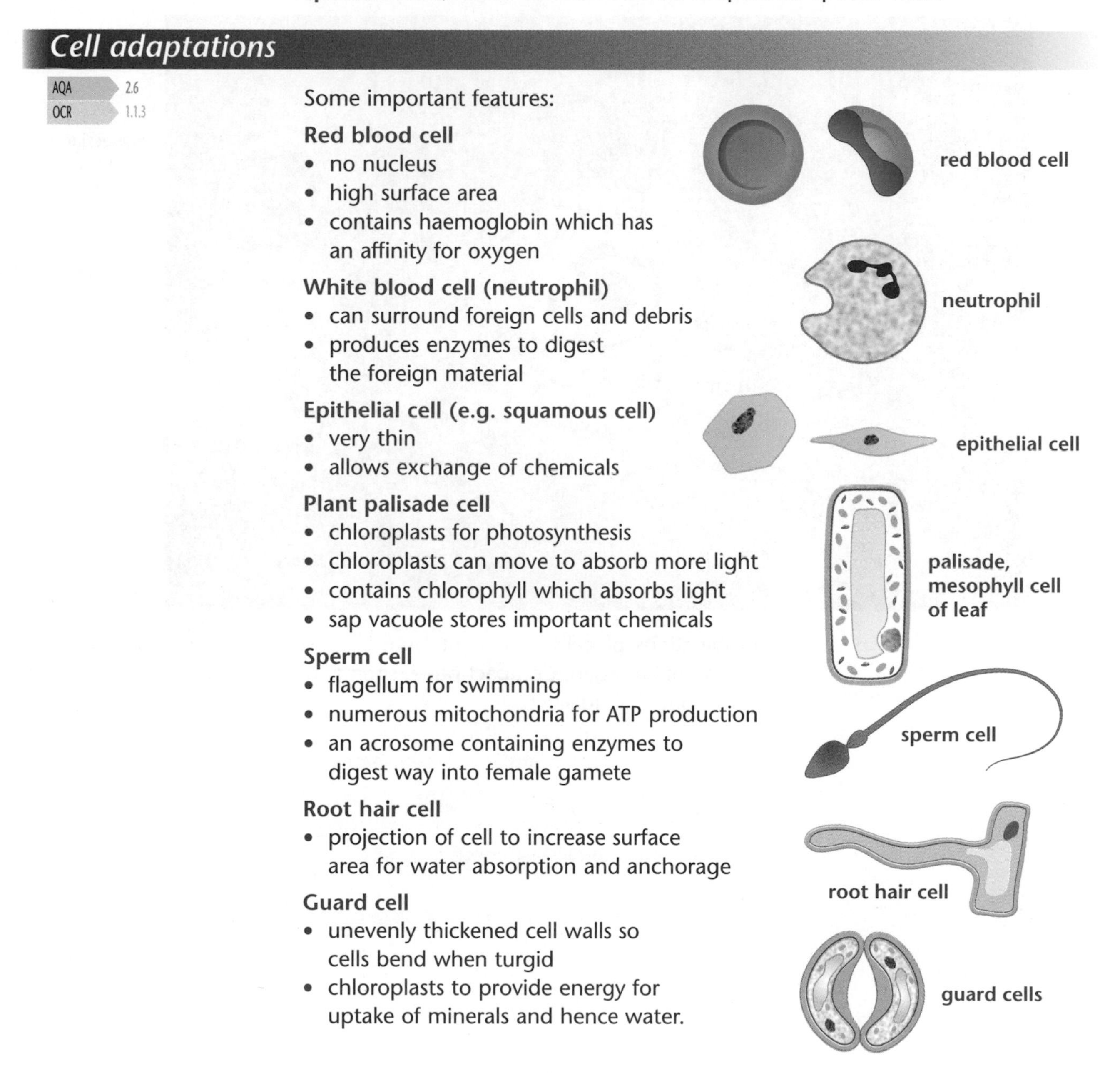

Some important features:

Red blood cell
- no nucleus
- high surface area
- contains haemoglobin which has an affinity for oxygen

White blood cell (neutrophil)
- can surround foreign cells and debris
- produces enzymes to digest the foreign material

Epithelial cell (e.g. squamous cell)
- very thin
- allows exchange of chemicals

Plant palisade cell
- chloroplasts for photosynthesis
- chloroplasts can move to absorb more light
- contains chlorophyll which absorbs light
- sap vacuole stores important chemicals

Sperm cell
- flagellum for swimming
- numerous mitochondria for ATP production
- an acrosome containing enzymes to digest way into female gamete

Root hair cell
- projection of cell to increase surface area for water absorption and anchorage

Guard cell
- unevenly thickened cell walls so cells bend when turgid
- chloroplasts to provide energy for uptake of minerals and hence water.

Tissues, organs and systems

AQA	2.6
CCEA	1.8
EDEXCEL	3.5
OCR	1.1.3
WJEC	1.2

A **tissue** is a collection of similar cells, derived from the same source, all working together for a specific function, e.g. palisade cells of the leaf which photosynthesise or the smooth muscle cells of the intestine which carry out peristalsis.

An **organ** is a collection of tissues which combine their properties for a specific function, e.g. the stomach includes: smooth muscle, epithelial lining cells, connective tissue, etc. Together they enable the stomach to digest food.

A range of tissues and organs combine to form a **system**, e.g. the respiratory system.

In multicellular organisms specific groups of cells are specialised for a particular role. This increased efficiency helps the organism to have better survival qualities in the environment.

The photomicrograph below shows some of the cells which are part of a bone.

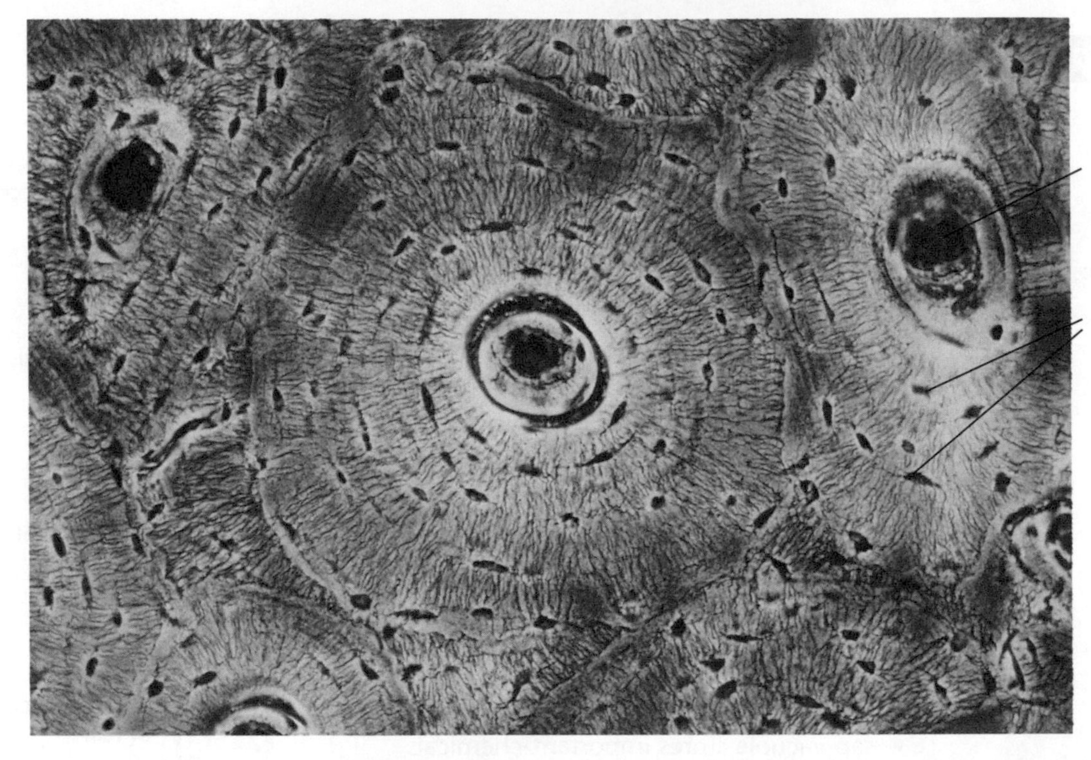

Combinations of cells each contribute their specific adaptations to the overall function of an organ. Compact bone, spongy bone and articular cartilage all have distinct but vital qualities.

Sample question and model answer

The electron micrograph below shows a lymphocyte which secretes antibodies. Antibodies are proteins.

Analyse electron micrographs carefully. Learn the typical structures of all the organelles and then you will be prepared.

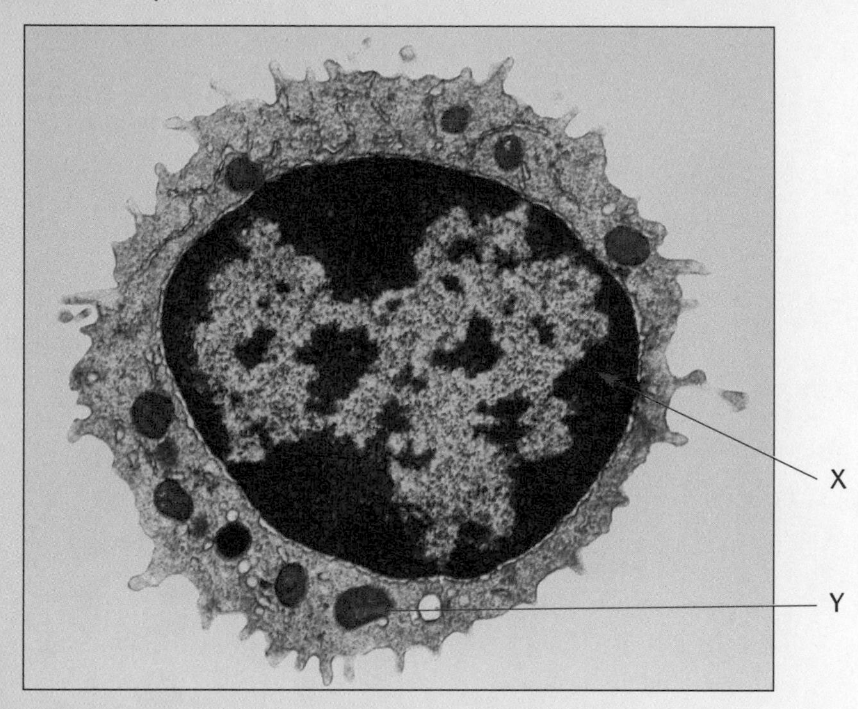

(a)

(i) Name the organelles X and Y. [2]

X = nucleus
Y = mitochondrion

(ii) The cell was stained with uranium salts in preparation for a transmission electron microscope. Explain how this stain caused the nucleus to show a dark shade compared to the light shade of the cytoplasm. [4]

There are many different stains. Most examination boards require that you know the principle of how they work, but not specific names.

The stain was taken up by the nucleus more than the cytoplasm; the electrons could not pass through the stained (dense) parts of the nucleus so the dark nucleus parts on the screen are in electron shade. Electrons pass through the cytoplasm and cause light emission (fluorescence) at the screen.

(b)

(i) Given a piece of liver how would you isolate mitochondria from the cells? [7]

Note that there are 7 marks maximum for this question. Any 7 of the responses shown would achieve maximum credit. The ideal answer given shows 9 potential creditable points.

Put the liver in isotonic solution;
homogenise the liver or grind with a pestle and mortar;
filter the homogenate through muslin layers to remove cell debris;
put in a centrifuge; spin at 500 – 600 g for 10 minutes;
discard the pellet or sediment; centrifuge the supernatant;
spin at 10 000 – 20 000 g for 20 minutes;
mitochondria are now in the pellet or sediment.

(ii) Why is it important to keep fresh liver cells at a temperature of around 5°C during the preparation of the sample? [2]

They keep alive for a longer period since a low temperature slows down the action of the enzymes which break down the mitochondria; i.e. prevents or slows down cell autolysis (self-digestion of the cells).

Practice examination questions

1 The diagram shows the structure of a cell surface membrane.

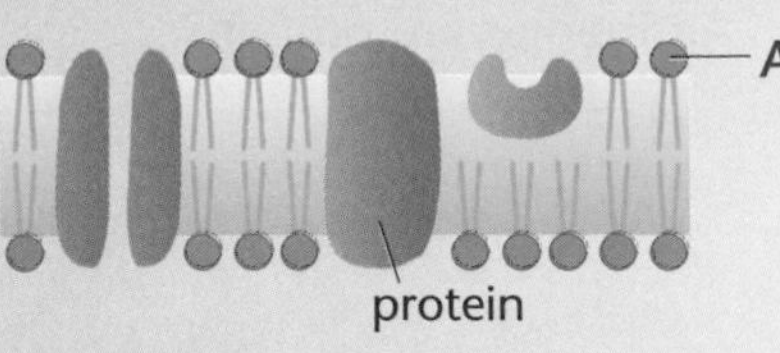

(a) Name molecule A. [1]

(b) Describe the role of protein molecules in:
(i) active transport
(ii) facilitated diffusion. [4]

2 The diagram below shows an electron micrograph of a cell.

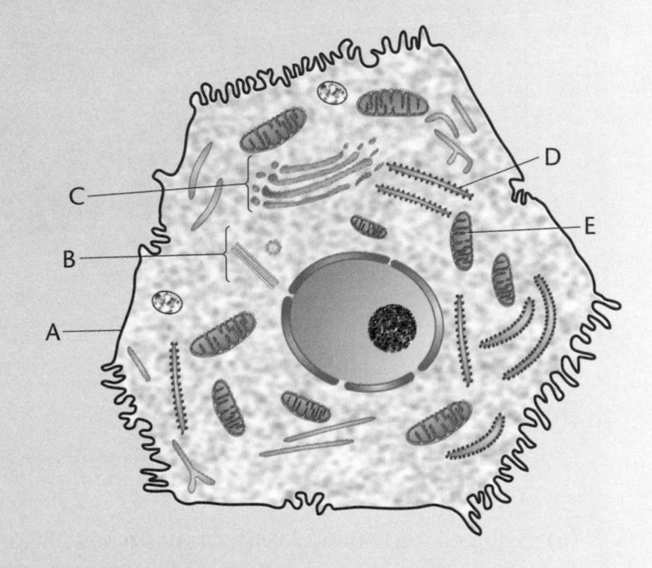

(a) Name the parts labelled in **A**, **B**, **C**, **D** and **E**. [5]

(b) The magnification of this diagram is 10 000.
Work out the actual diameter of the nucleus.
Give your answer in micrometers. [3]

(c) The cell is a liver cell. It contains many mitochondria.
Explain why there are so many mitochondria in each liver cell. [2]

3 The diagram below shows a bacterium.

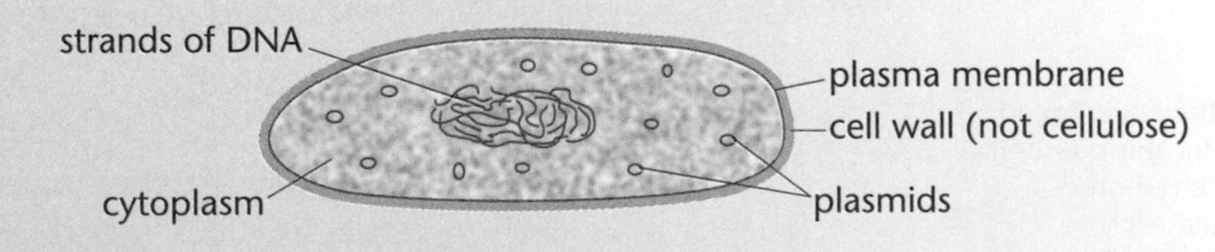

(a) Describe **two** features, visible in the diagram, which show that the bacterium is a prokaryotic organism. [2]

(b) Name **two** organelles from a human cell which show that it is a eukaryotic organism. [2]

Practice examination questions (continued)

4 The diagram below shows the structure of a transmission electron microscope (TEM).

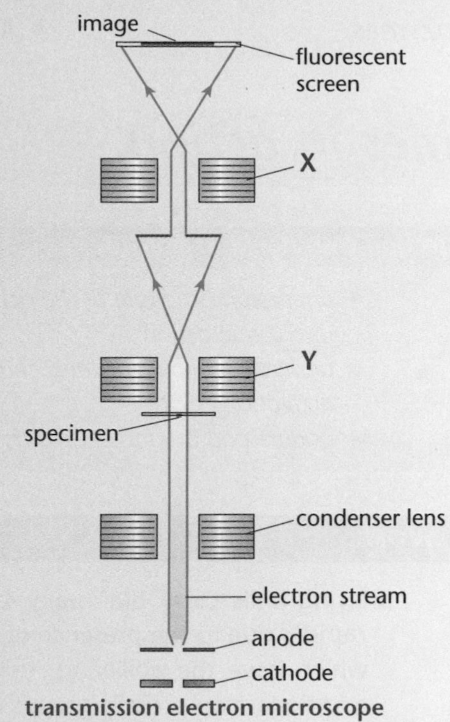

transmission electron microscope

(a) Name lens X and lens Y. [2]

(b) Why is it necessary for the specimen to be put in a vacuum? [1]

(c) Occasionally an image seen when using the electron microscope shows an item not present in the living organism.

(i) What name is given to this type of item? [1]

(ii) How should the presence of the item be interpreted? [1]

Chapter 3

Enzymes

The following topics are covered in this chapter:

- *Enzymes in action*
- *Inhibition of enzymes*
- *Enzymes and digestion*
- *Applications of enzymes*

3.1 Enzymes in action

After studying this section you should be able to:

LEARNING SUMMARY

- *understand the role of the active site and the enzyme–substrate complex in enzyme action*
- *understand how enzymes catalyse biochemical reactions by lowering activation energy*
- *understand the factors which affect the rate of enzyme catalysed reactions*

How enzymes work

AQA	3.1.1
CCEA	1.2
EDEXCEL	2.8–9
OCR	2.1.3
WJEC	1.4

Living cells carry out many biochemical reactions. These reactions take place **rapidly** due to the presence of enzymes. All enzymes consist of **globular proteins** which have the ability to 'drive' biochemical reactions. Some enzymes require additional non-protein groups to enable them to work efficiently, e.g. the enzyme dehydrogenase needs a coenzyme NAD to function. Most enzymes are contained within cells but some may be released and act extracellularly.

The tertiary folding of polypeptides are responsible for the special shape of the active site.

KEY POINT

The ability of an enzyme to function depends on the specific shape of the protein molecule. The intricate shape created by polypeptide folding (see page 27) is a key factor in both theories of enzyme action.

Lock and key theory

In an examination the lock and key theory is the most important model to consider. Remember that both catabolic and anabolic reactions may be given.

- Some part of the enzyme has a cavity with a precise shape (**active site**).
- A substrate can fit into the active site.
- The active site (lock) is exactly the correct shape to fit the substrate (key).
- The substrate binds to the enzyme forming an **enzyme–substrate complex**.
- The reaction takes place rapidly.
- Certain enzymes break a substrate down into two or more products (**catabolic** reaction).
- Other enzymes bind two or more substrates together to assemble one product (**anabolic** reaction).

metabolic reaction = anabolic reaction + catabolic reaction

Remember that metabolism is a summary of **build up** and **break down reactions.**

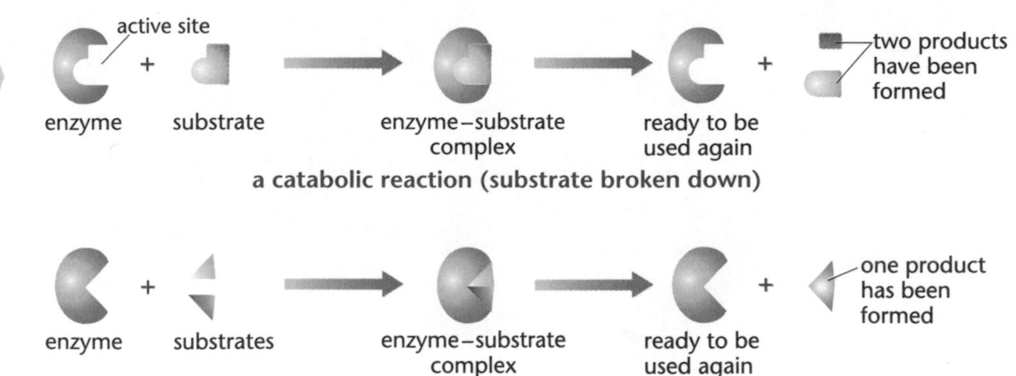

Induced fit theory

- The active site is a cavity of a particular shape.
- Initially the active site is not the correct shape in which to fit the substrate.
- As the substrate approaches the active site, the site changes and this results in it being a perfect fit.
- After the reaction has taken place, and the products have gone, the active site returns to its normal shape.

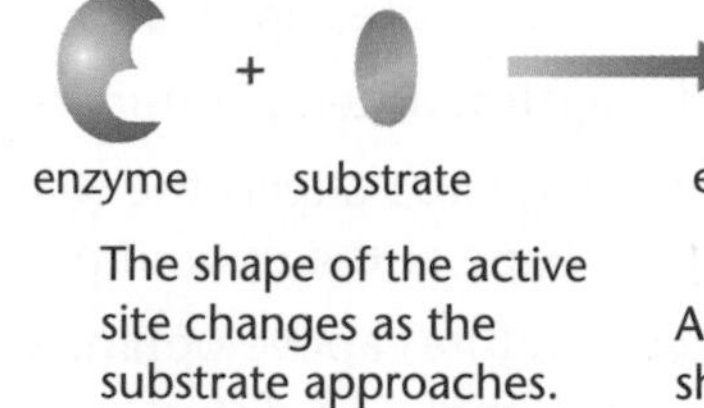

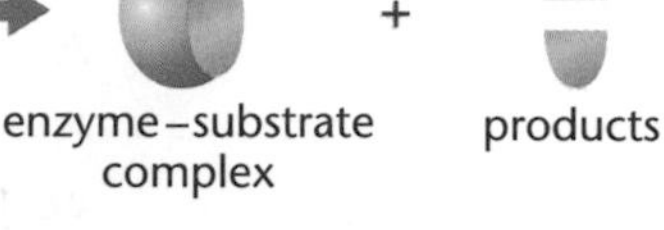

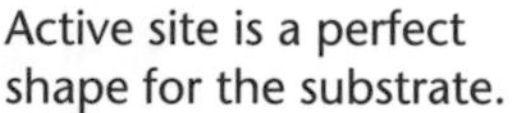

The shape of the active site changes as the substrate approaches.

Active site is a perfect shape for the substrate.

Lowering of activation energy

Every reaction requires the input of energy. Enzymes reduce the level of activation energy needed as shown by the graph.

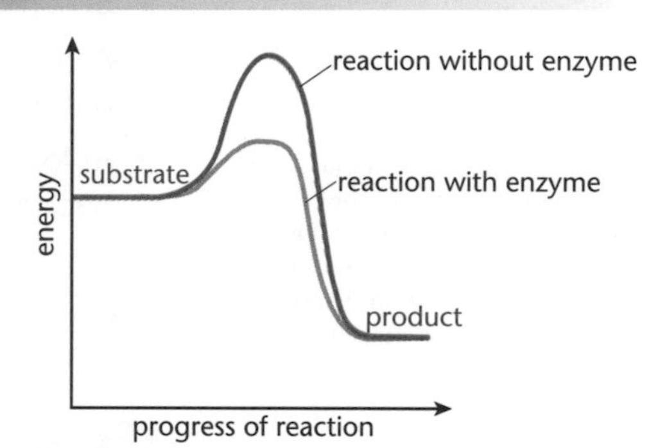

The higher the activation energy the slower the reaction. An enzyme reduces the amount of energy required for a biochemical reaction. When an enzyme binds with a substrate the available energy has a greater effect and the rate of catalysis increases. The conditions which exist during a reaction are very important when considering the rate of progress. Each of the following has an effect on the rate:

- concentration of substrate molecules
- concentration of enzyme molecules
- temperature
- pH.

You may be questioned on the factors which affect the rate of reaction. Less able candidates tend to remember just one or two factors. Learn all four factors here and achieve a higher grade!

What is the effect of enzyme and substrate concentration?

When considering the rate of an enzyme catalysed reaction the proportion of enzyme to substrate molecules should be considered. Every substrate molecule fits into an active site, then the reaction takes place. If there are more substrate molecules than enzyme molecules then the number of active sites available is a **limiting factor**. The **optimum rate** of reaction is achieved when all the active sites are in use. At this stage if more substrate is added, there is no increase in rate of product formation. When there are fewer substrate molecules than enzyme molecules the reaction will take place very quickly, as long as the conditions are appropriate.

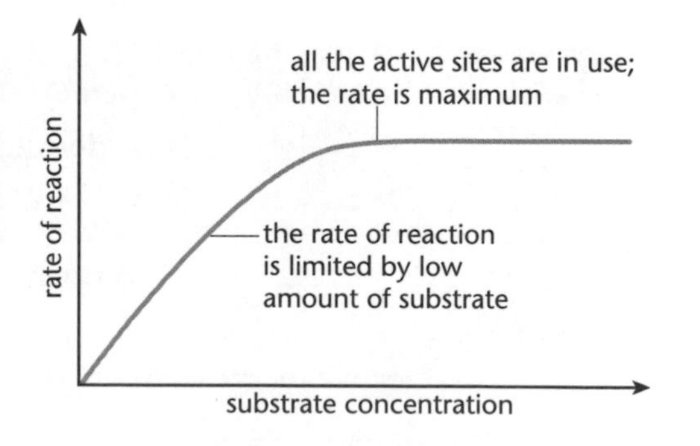

Look out for questions which show the rate of reaction graphically. The examiners often test your understanding of limiting factors (see practice question on page 54).

Remember that particles in liquids (and gases) are in constant random motion, even though we cannot see them.

How does temperature affect the rate of an enzyme catalysed reaction?

- **Heat energy** reaching the enzyme and substrate molecules causes them to **increase random movement.**
- The greater the heat energy the more the molecules move and so the more often they **collide**.
- The more **collisions** there are the greater the chance that substrates will fit into an **active site**, up to a specific temperature.
- At the **optimum** temperature of an enzyme, the reaction rate is maximum.
- Heat energy also affects the shape of the active site, the active site being the correct shape at the optimum temperature.
- At temperatures above optimum, the rate of reaction decreases because the active site begins to distort.
- Very high temperature causes the enzyme to become **denatured**, i.e. bonding becomes irreversibly changed and the active site is **permanently damaged**.
- At very high temperatures, the number of collisions is correspondingly high, but without active sites no products can be formed.
- At lower temperatures than the optimum, the rate of the reaction decreases because of reduced enzyme–substrate collisions.

It is interesting to consider that some micro organisms can spoil ice-cream in a freezer whereas a different micro organism, with different enzymes, can decompose grass in a 'steaming' compost heap.

Most enzymes have an optimum temperature of between 30°C and 40°C, but there are many exceptions. An example of this is shown by some bacteria that live at high temperatures in hot volcanic springs.

How does pH affect the rate of an enzyme catalysed reaction?

The **pH** of the medium can have a direct effect on the bonding responsible for the **secondary and tertiary structure** of enzymes. If the active site is changed then enzyme action will be affected. Each enzyme has an optimum pH.

Remember that other factors affect an enzyme catalysed reaction:

- substrate concentration
- enzyme concentration
- temperature.

Each can be a limiting factor.

- Many enzymes work best at **neutral** or **slightly alkaline** conditions, e.g. salivary amylase.
- Pepsin works best in **acid** conditions around pH 3.0, as expected considering that the stomach contains hydrochloric acid.

For the two enzyme examples above, the active sites are ideally shaped at the pHs mentioned. An inappropriate pH, often acidic, can change the active site drastically, so that the substrate cannot bind. The reaction will not take place. On most occasions the change of shape is not permanent and can be returned to optimum by the addition of an alkali.

Progress check

How does temperature affect the rate of the reaction by which protein is changed to polypeptides by the enzyme pepsin, in the human stomach?

$$\text{protein} \xrightarrow{\text{pepsin}} \text{polypeptides}$$

- Heat energy causes the enzyme and substrate molecules to increase random movement, increasing the chance of collision.
- At 37°C (optimum temperature) there is a greater chance that the protein will fit into an active site, so the production of polypeptides is at maximum rate.
- At 37°C (optimum temperature) the shape of the active site is best suited to fit the protein.
- At temperatures higher than 37°C the rate of reaction decreases because the active site begins to distort.
- Very high temperature causes the pepsin to become denatured, i.e. bonding has been irreversibly changed and the active site is permanently damaged.
- At very high temperatures the number of collisions is correspondingly high, but without active sites no polypeptides can be formed.
- At lower temperatures than optimum the rate of reaction decreases because of reduced enzyme–substrate collisions.

3.2 Inhibition of enzymes

After studying this section you should be able to:

- *understand the action and effects of competitive and non-competitive inhibitors*

LEARNING SUMMARY

What are inhibitors?

AQA	3.1.1
CCEA	1.2
OCR	2.1.3
WJEC	1.4

Certain chemicals can slow down or stop enzyme catalysed reactions. These chemicals are called **inhibitors**.

Sometimes these chemicals are substances that are naturally occurring inside cells. They may be used to regulate the rate of enzyme controlled reactions. Other inhibitors may be poisons or medicines.

Competitive inhibitors

- These are molecules of **similar shape** to the normal substrate and are able to bind to the active site.
- They do not react within the active site, but leave after a time without any product forming.
- The enzymic reaction is **reduced** because while the inhibitor is in the active site, **no substrate can enter**.
- Substrate molecules **compete** for the active site so the rate of reaction decreases.
- The higher the proportion of competitive inhibitor the slower the rate of reaction.

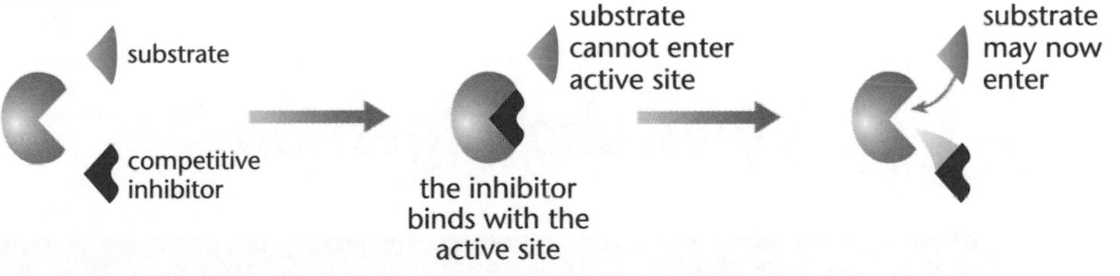

Some enzymes have two sites, the active site and one other. An **allosteric** molecule fits into the alternative site. Here it changes the shape of the active site. This can stimulate the reaction if the active site becomes a better shape **(allosteric activation)**. It can also inhibit if the active site becomes an inappropriate shape **(allosteric inhibition)**.

Non-competitive inhibitors

- These are molecules which bind to some part of an enzyme other than the active site.
- They have a different shape to the normal substrate.
- They change the shape of the active site which no longer allows binding of the substrate.
- Some substrate molecules may reach the active site before the non-competitive inhibitor.
- The rate of reaction is reduced.
- Finally they leave their binding sites, but substrate molecules do not compete for these, so they have a greater inhibitory effect.

non-competitive inhibitor
substrate
binding site
active site has changed
substrate has opportunity to enter

The graph below shows the relative effects of competitive and non-competitive inhibitors, compared to a normal enzyme catalysed reaction.

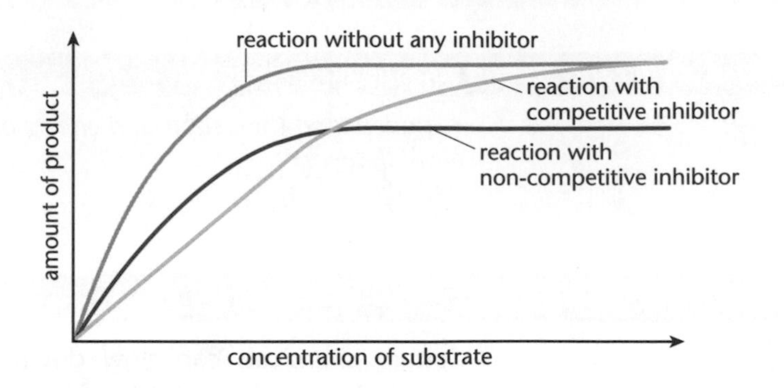

Sometimes an inhibitor will not leave the enzyme once it has bound with it. It stays permanently attached. This is called an **irreversible inhibitor**. Inhibitors that do leave are called **reversible**.

Progress check

1 Explain why a non-competitive inhibitor does not need to have the same shape as the substrate. Give a reason for your answer.

2 How does a non-competitive inhibitor reduce the rate of an enzyme catalysed reaction?

1 No. Non-competitive inhibitors have a different shape to the normal substrate. They bind to some part of an enzyme other than the active site.

2 They change the shape of the enzyme's active site which is less suitable for the binding of the substrate, so the rate of reaction decreases.

3.3 Enzymes and digestion

After studying this section you should be able to:

- *recall the structure of the human digestive system*
- *recall the sites of secretion and action of enzymes involved in carbohydrate digestion*

LEARNING SUMMARY

The human digestive system

AQA 1.2
CCEA 1.8.2
WJEC 2.5

Humans feed **heterotrophically** by taking in complex organic substances. This is called **ingestion**. These substances are then broken down into smaller, soluble molecules by **digestion**.

Once digestion has occurred, the molecules can be **absorbed** into the blood stream. They can then be built up or **assimilated** into the various complex molecules that the body requires.

The human digestive system is made up of a long tube that starts at the mouth and ends at the anus. Along its length, different secretions are added to the food either from the lining of the system or from associated glands.

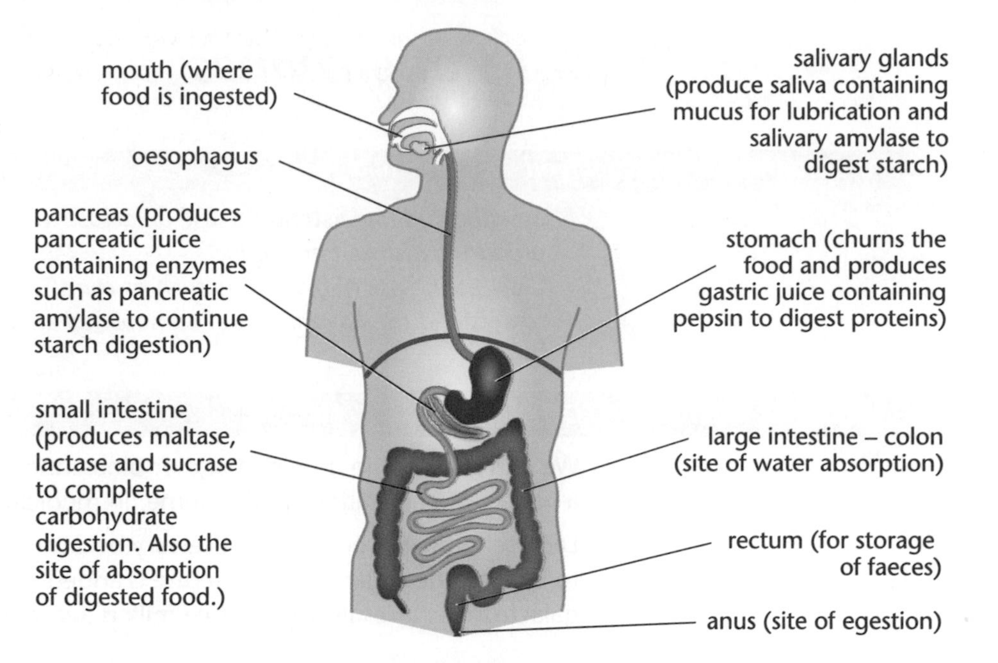

Although different regions of the digestive system are specialised for different jobs, they are all built on a similar pattern. The diagram shows the different layers in the wall of the digestive system.

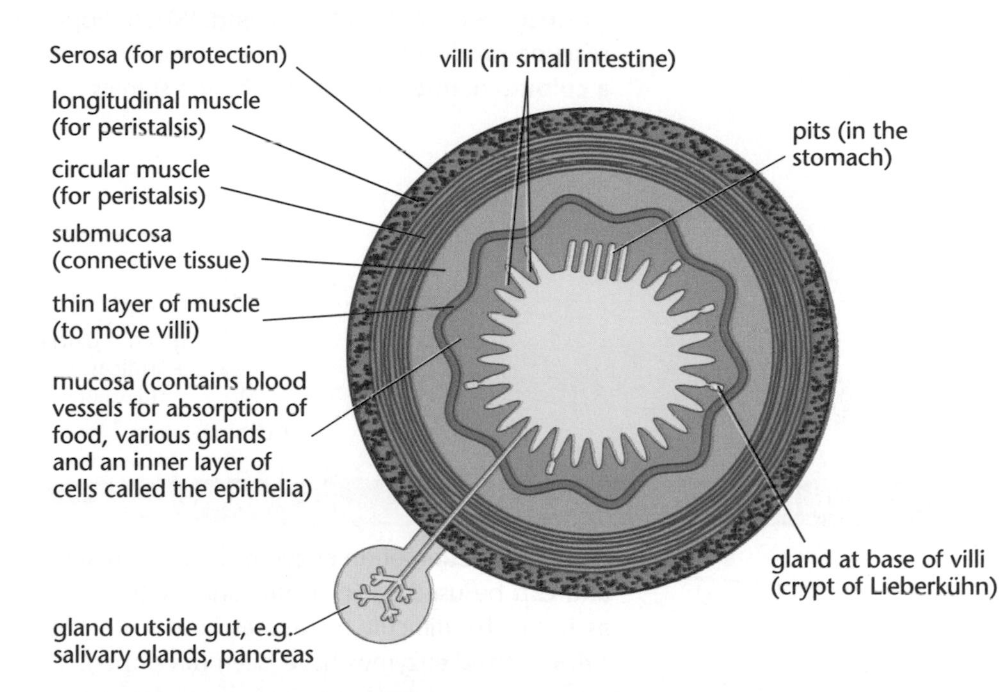

Digestion of carbohydrates

AQA 1.2

The human digestive enzymes are **hydrolases**. Each time a molecule is broken down water is required during the reaction. The table below includes some human digestive enzymes that are involved in carbohydrate digestion.

Site in alimentary canal	*Secretion*	*Enzyme*	*Substrate*	*Product*
mouth	saliva	amylase	starch	maltose
duodenum	pancreatic juice	amylase	starch	maltose
small intestine	intestinal juice	maltase	maltose	glucose
		sucrase	sucrose	glucose & fructose
		lactose	lactose	glucose & galactose

3.4 Applications of enzymes

After studying this section you should be able to:

- *describe and understand the use of lactase in producing lactose-free products*
- *describe the use of biosensors such as clinistix*
- *understand the advantages of immobilised enzymes*

LEARNING SUMMARY

Lactose removal from milk

AQA 1.2

We currently live in a golden age of biotechnology and enzymes are now used in a variety of applications in the home, in medicine and in industry.

Lactose removal from milk is necessary because some people have allergic reactions to lactose, the sugar in milk. **Lactase** is added to change the lactose to glucose and galactose. As a result, lactose-free milk is available to consumers.

Biosensors

CCEA 1.2.3
WJEC 1.5

There are a number of reagent strips that use enzymes to detect various chemicals. An example is clinistix strips. These are strips of cellulose which have the enzyme **glucose oxidase** stuck to one end. When dipped into urine containing glucose the reaction produces **hydrogen peroxide**. This reacts with another compound to give a **colour change**. This colour change signals that the patient is potentially diabetic.

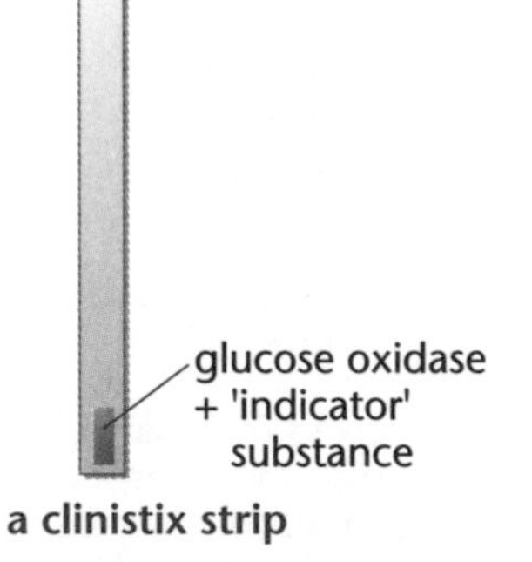

a clinistix strip

Immobilised enzymes

CCEA 1.2.3
WJEC 1.5

After completion of an enzyme catalysed reaction the enzyme remains unchanged and can be used again. Unfortunately the enzyme can contaminate the product, as it can be difficult to separate out from the reaction mixture. For this reason **immobilised** enzymes have been developed. They are used as follows:

- enzymes are attached to insoluble substances such as resins and alginates
- these substances usually form membranes or beads and the enzymes bind to the outside
- substrate molecules readily bind with the active sites and the normal reactions go ahead
- the immobilised enzymes are easy to recover, remaining in the membranes or beads
- there is no contamination of the product by free enzymes
- expensive enzymes are re-used
- processes can be continuous unlike batch, where the process is stopped for 'harvesting'.

Sample question and model answer

The graph below shows the rate of an enzyme catalysed reaction, with and without a **non-competitive inhibitor**, at different substrate concentrations.

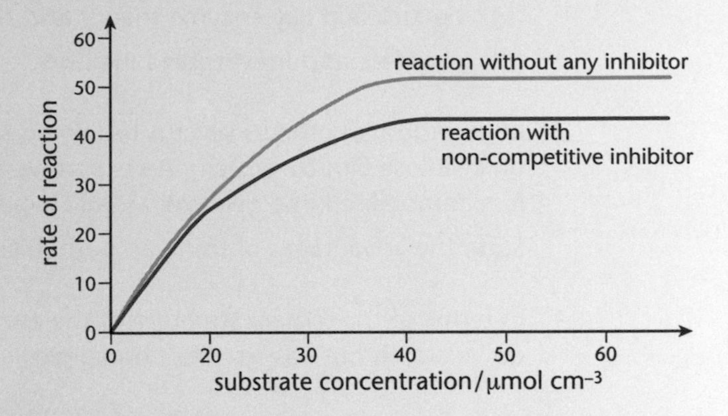

(a) Give **one** piece of evidence from the graph which shows that the inhibitor was non-competitive rather than competitive. [1]

When a non-competitive inhibitor is present the maximum rate of the inhibitor-free reaction is not reached.

A competitive inhibitor would still allow the maximum rate at high substrate concentration.

(b) Explain why the rates of the inhibited and non-inhibited reactions were very similar up to a substrate concentration of 20 µmol cm^{-3}. [2]

Before a concentration of 20 µmol cm^{-3}
- *in both reactions substrate molecules are similarly successful in reaching active sites*
- *few inhibitor molecules have become bound to the alternative sites on the enzyme*
- *few active sites have been changed so are still available for the substrate.*

Remember that the non-competitive inhibitor binds to a different part of an enzyme and NOT the active site. It still causes a change in the active site which cannot then bind with the substrate.

(c) Give **two** factors that are needed to be kept constant when investigating both the inhibited and non-inhibited reactions. [2]

- *temperature*
- *pH*

Practice examination questions

1 Describe each of the following pairs to show that you understand the main differences between them:

(a) the lock and key enzyme theory **and** the induced fit enzyme theory

(b) reversible **and** irreversible inhibitors. [4]

2 In an industrial process silver is reclaimed from a waste fluid. The silver is held on cellulose film by protein. An expensive enzyme is used to remove the protein. It is immobilised in a gel layer as fluid is passed over it.

State the advantages of using an immobilised enzyme in this process. [2]

3 In terms of the tertiary structure of the enzyme, explain why amylase can break down starch but has no effect on a lipid. [3]

4 Bacterial α-amylase works best at around 80°C.

(a) Name the substrate which it breaks down. [1]

(b) Why is this enzyme described as a thermostable enzyme? [1]

5 The diagram represents an enzyme and its substrate.

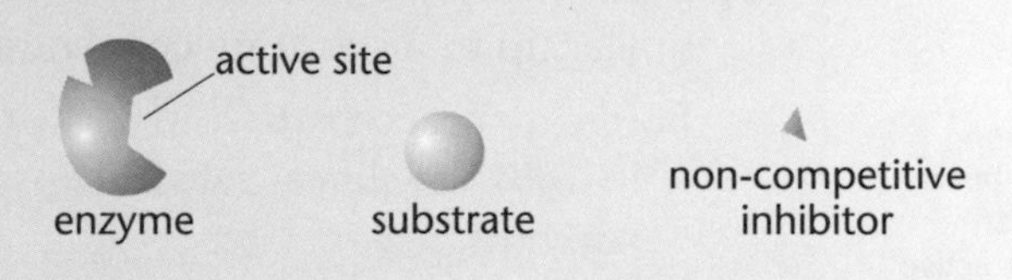

(a) Referring to information in the diagram explain the activity of this enzyme in terms of the **induced fit theory**. [2]

(b) Molecule X is a non-competitive inhibitor. Explain how this inhibitor has an effect on the structure and function of the enzyme. [3]

6 The diagram below shows a long polypeptide.

(a) The carboxylic acid is found at one end of the polypeptide. Which group is found at the other end? [1]

(b) Pepsin breaks down polypeptides by breaking bonds in the centre of the molecule. This speeds up the subsequent digestion of the polypeptide. Explain how. [2]

Chapter 4

Exchange

The following topics are covered in this chapter:

- *The cell surface membrane*
- *The movement of molecules in and out of cells*
- *Gaseous exchange in humans*
- *Gaseous exchange in other organisms*

4.1 The cell surface membrane

After studying this section you should be able to:

- *understand the importance of surface area to volume ratio*
- *recall the fluid mosaic model of the cell surface (plasma) membrane*

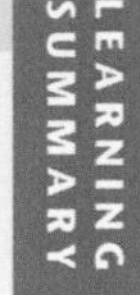

How important is the surface area of exchange surfaces?

AQA A	2.7
CCEA	2.1
OCR	1.2.1
WJEC	2.2

Unicellular organisms like amoeba have a very high **surface area to volume ratio.** All chemicals that are needed can pass into the cells directly and all waste can pass out efficiently. Organisms which have a high surface area to volume ratio have no need for special structures like lungs or gills.

Nutrients and oxygen passing into an organism are rapidly used up. This limits the ultimate size to which a unicellular organism can grow. If vital chemicals did not reach all parts of a cell then death would be a consequence.

A unicellular organism may satisfy all its needs by direct diffusion. However, in larger organisms cells join to adjacent ones and surfaces exposed for exchange of substances are reduced. The larger an organism the lower its surface area to volume ratio. For this reason many multicellular organisms have specially adapted exchange structures.

Fluid mosaic model of the cell surface (plasma) membrane

AQA	1.3
CCEA	1.5
EDEXCEL	2.2
OCR	1.1.2
WJEC	1.3

Ultimately the exchange of substances takes place across the cell surface membrane. This must be selective, allowing some substances in and excluding others. The cell membrane consists of a bilayer of phospholipid molecules (see page 25). Each phospholipid is arranged so that the hydrophilic (attracts water) head is facing towards either the cytoplasm or the outside of the cell. The hydrophobic (repels water) tails meet in the middle of the membrane. Across this expanse of phospholipids are a number of protein molecules. Some of the proteins (intrinsic) span the complete width of the membrane, some proteins (extrinsic) are partially embedded in the membrane.

Remember that plant cells have a cellulose cell wall. This gives physical support to the cell but is permeable to many molecules. Water and ions can readily pass through.

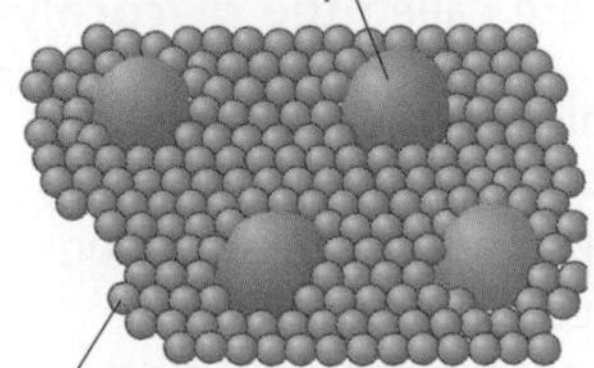

The fluid mosaic model of the cell membrane

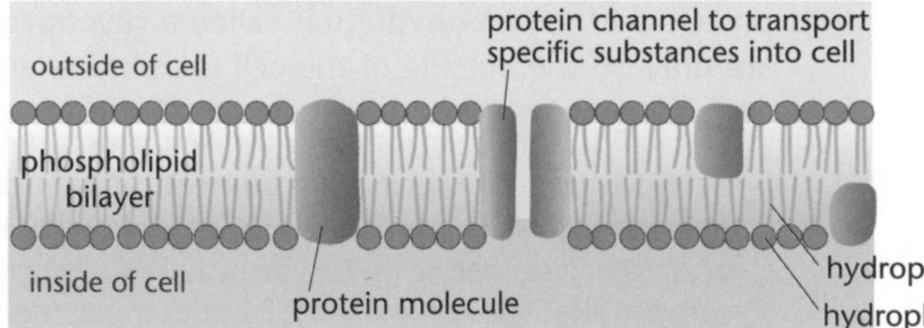

Functions of cell membrane molecules

The term 'fluid mosaic' was given to the cell membrane because of the dynamic nature of the component molecules of the membrane. Many of the proteins seem to 'float' through an apparent 'sea' of phospholipids. Few molecules are static. The fluidity of the membrane is controlled by the quantity of cholesterol molecules. These are found inbetween the tails of the phospholipids.

Phospholipid

Small lipid-soluble molecules pass through the membrane easily because they dissolve as they pass through the phospholipid's bilayer. Small uncharged molecules also pass through the bilayer.

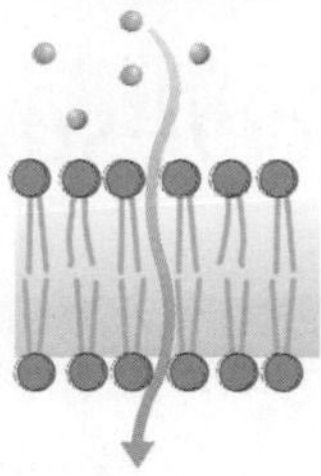

small lipid-soluble molecules pass through

Channel proteins (ion gates)

Larger molecules and charged molecules can pass through the membrane due to channel proteins. Some are adjacent to a receptor protein, e.g. at a synapse a transmitter substance binds to a receptor protein. This opens the channel protein or ion gate and sodium ions flow in.

Not all channel proteins need a receptor protein.

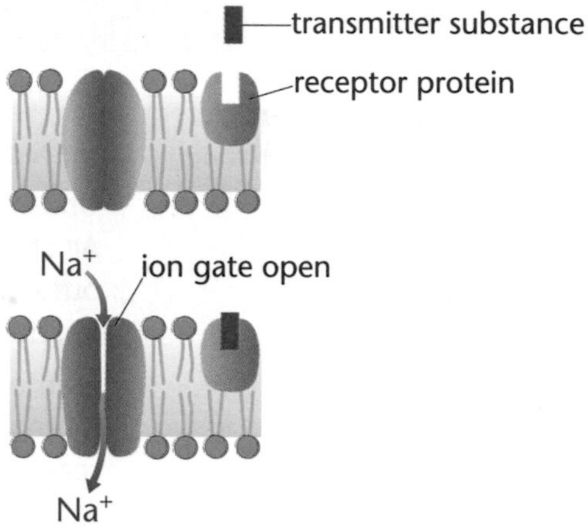

transmitter substance

receptor protein

Na^+ ion gate open

Na^+

When the molecule binds to a receptor molecule it is similar to a substrate binding with an enzyme's active site. On this occasion the receptor site is the correct shape.

Carrier protein molecule

Some molecules which approach a cell may bind with a carrier protein. This has a site which the incoming molecule can bind to. This causes a change of shape in the carrier protein which deposits the molecule into the cell cytoplasm.

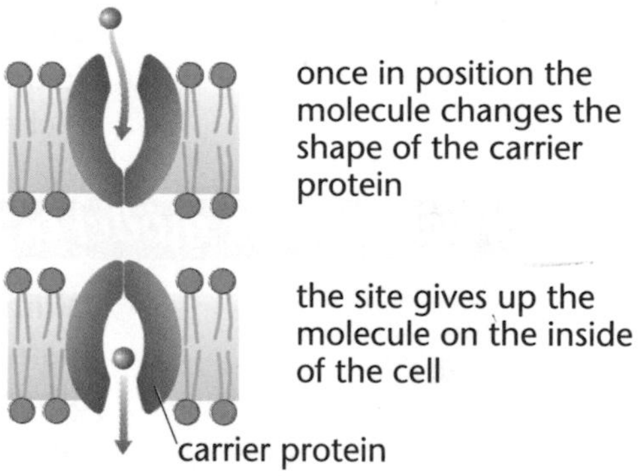

once in position the molecule changes the shape of the carrier protein

the site gives up the molecule on the inside of the cell

carrier protein

Recognition proteins

These are extrinsic proteins, some having carbohydrate components, which help in cell recognition (cell signalling) and cell interaction, e.g. foreign protein on a bacterium would be recognised by white blood cells and the cell would be attacked. The combination of a protein with a carbohydrate is called a **glycoprotein.** The carbohydrate chains are only on the outside of the cell membrane and are called the **glycocalyx.**

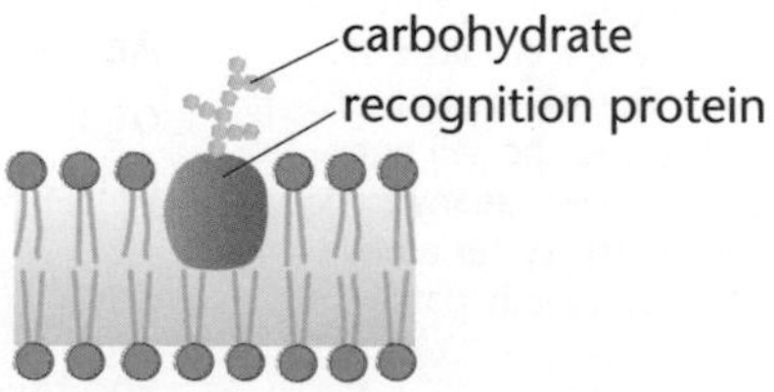

carbohydrate

recognition protein

White blood cells continually check the proteins on cell membranes. Those recognised as 'self' are not attacked, whereas those which are not 'self' are attacked.

KEY POINT

The cell surface membrane is the key structure which forms a barrier between the cell and its environment. Nutrients, water and ions must enter and waste molecules must leave. Equally important is the exclusion of dangerous chemicals and inclusion of vital cell contents. It is no surprise that the cell makes further compartments within the cell using membranes of similar structure to the cell surface membrane. High temperatures can destroy the structure of the cell membrane and then it loses its ability to contain the cell contents.

4.2 The movement of molecules in and out of cells

After studying this section you should be able to:

LEARNING SUMMARY

- *understand the range of methods by which molecules cross cell membranes*
- *understand the processes of diffusion, facilitated diffusion, osmosis and active transport*

How do substances cross the cell surface membrane?

AQA	1.3
CCEA	1.6
EDEXCEL	2.3-5
OCR	1.1.2
WJEC	1.3

Cells need to obtain substances vital in sustaining life. Some cells secrete useful substances but all cells excrete waste substances. There are several mechanisms by which molecules move across the cell surface membrane.

Diffusion

Note that diffusion is the movement of molecules down a concentration gradient.

Molecules in liquids and gases are in constant random motion. When different concentrations are in contact, the molecules move so that they are in equal concentration throughout. An example of this is when sugar is put into a cup of tea. If left, sugar molecules will distribute themselves evenly, even without stirring. Diffusion is the movement of molecules from where they are in high concentration to where they are in low concentration. Once evenly distributed the *net* movement of molecules stops.

Factors which affect the rate of diffusion

- Surface area.
  the greater the surface area the greater the rate of diffusion
- The difference in concentration on either side of the membrane.
  the greater the difference the greater the rate
- The size of molecules.
  smaller molecules may pass through the membrane faster than larger ones
- The presence of pores in the membrane.
  pores can speed up diffusion
- The width of the membrane.
  the thinner the membrane the faster the rate.

Sometimes the membrane is stated as being selectively permeable, partially permeable or semi-permeable. It depends on which examination board sets your papers.

Facilitated diffusion

This is a special form of diffusion in which protein carrier molecules are involved. It is much faster than regular diffusion because of the carrier molecules. Each carrier will only bind with a specific molecule. Binding changes the shape of the carrier which then deposits the molecule into the cytoplasm. No energy is used in the process.

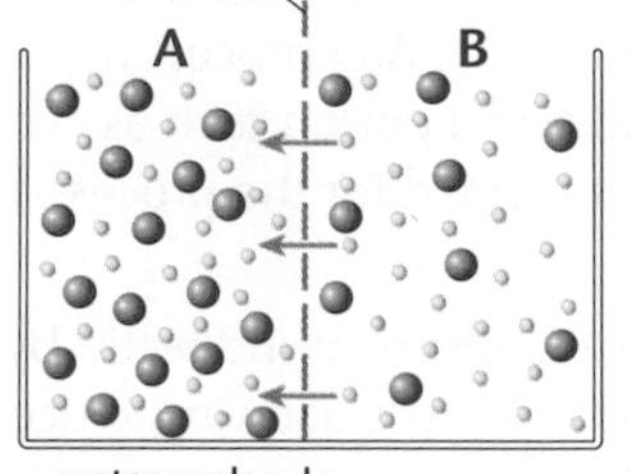

water molecules move from B to A

Osmosis

This is the movement of water molecules across a selectively permeable membrane:

- from a lower concentrated solution to a higher concentrated solution
- from where water molecules are at a higher concentration to where they are at a lower concentration
- from a hypotonic solution to a hypertonic solution
- from a hyperosmotic solution to a hypo-osmotic solution
- from an area of higher water potential to lower water potential.

The diagram on the left shows a model of osmosis.

Remember that osmosis is about the movement of **water molecules**. No other substance moves!

What is the relationship between water potential of the cell and the concentration of an external solution?

The term 'water potential' is used as a measure of water movement from one place to another in a plant. It is measured in terms of pressure and the units are either kPa (kilopascals) or MPa (megapascals). Water potential is indicated by the symbol ψ *(pronounced psi)*. The following equation allows us to work out the water 'status' of a plant cell.

ψ (cell) water potential (of cell)	=	ψs solute potential (of ions inside cell)	+	ψp pressure potential (of cell wall)	KEY POINT

Remember that water moves from an area of higher water potential to an area of lower water potential. When a cell at –4 MPa is next to a cell at the less negative value the water moves to the more negative value, i.e. –4 MPa > –6 MPa

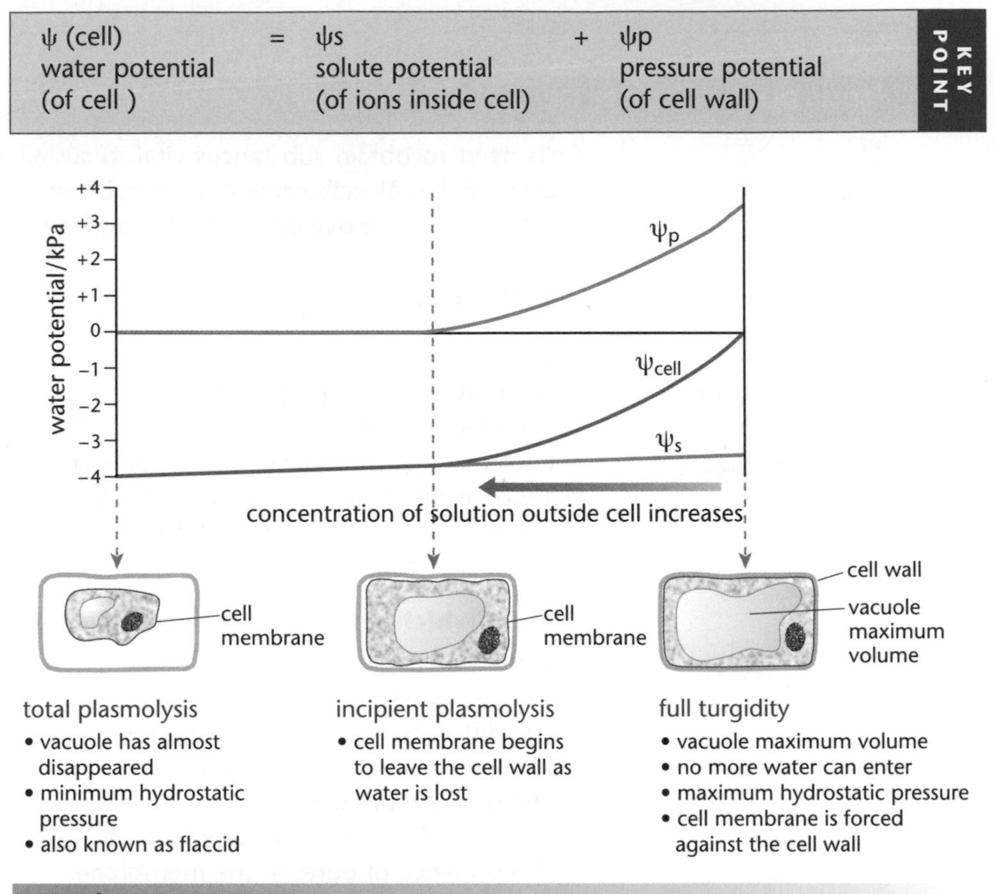

total plasmolysis
- vacuole has almost disappeared
- minimum hydrostatic pressure
- also known as flaccid

incipient plasmolysis
- cell membrane begins to leave the cell wall as water is lost

full turgidity
- vacuole maximum volume
- no more water can enter
- maximum hydrostatic pressure
- cell membrane is forced against the cell wall

Note that pressure potential only has a value **above** zero when the cell membrane **begins** to contact the cell wall. The greater the pressure potential the more the cell wall resists water entry. At turgidity $\psi s = \psi p$ when net water movement is zero.

Active transport

In **active transport** molecules move from where they are in lower concentration to where they are in higher concentration. A protein carrier molecule is used. This is **against the concentration gradient** and always **needs energy**. A plant may contain a higher concentration of Mg^{2+} ions than the soil. It obtains a supply by active transport through the cell surface membranes of root hairs. Only Mg^{2+} ions can bind with the specific protein carrier molecules responsible for their entry into the plant. This is also known as active ion uptake, but is a form of active transport. Any process that reduces respiration in cells will reduce active transport, e.g. adding a poison such as cyanide or reducing oxygen availability. This is because energy is needed for the process and this energy is released by respiration.

Note that active transport is the movement of molecules up a concentration gradient.

In the small intestine, glucose is absorbed into the bloodstream indirectly by active transport. Sodium ions are pumped out of the epithelial cells allowing glucose and sodium ions to diffuse into the cell by a co-transporter system.

endocytosis

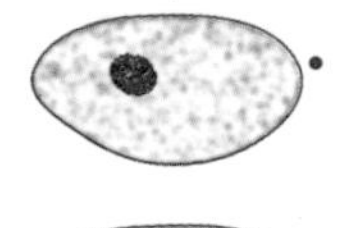

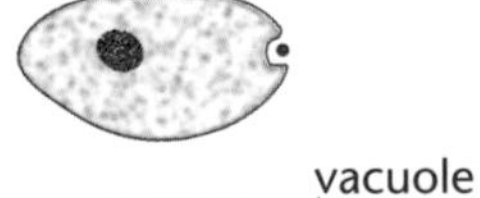

Endocytosis, exocytosis, pinocytosis and phagocytosis

Some substances, often due to their large size, enter cells by **endocytosis** as follows:

- the substance contacts the cell surface membrane which indents
- the substance is surrounded by the membrane, forming a vacuole or vesicle
- each vacuole contains the substance and an outer membrane which has detached from the cell surface membrane.

When fluids enter the cell in this way this is known as **pinocytosis**. When the substances are large solid particles, this is called **phagocytosis**. Some substances leave the cell in a reverse of endocytosis. Here the membrane of the vacuole or vesicle merges with the cell surface membrane depositing its contents into the outside environment of the cell. This is known as **exocytosis**.

Progress check

1 A plant contains a greater concentration of Fe^{2+} ions than the soil in which it is growing. Name and describe the process by which the plant absorbs the ions against the concentration gradient.

2 Explain the following:
(a) Endocytosis of an antigen by a phagocyte
(b) Exocytosis of amylase molecules from a cell.

1 **Active transport:** molecules move from a lower concentration to a higher concentration; through the cell surface membranes of root hairs; protein carrier molecules in membranes used; energy needed; Fe^{2+} ions can bind with the protein carrier molecules which allow entry into the plant.

2 (a) **Endocytosis:** antigen contacts the cell membrane of the phagocyte; cell membrane surrounds the antigen, forming a vacuole; the vacuole contains the antigen and an outer membrane which has detached from the cell surface membrane.
(b) **Exocytosis:** a vesicle in the cell contains amylase molecules; the vesicle merges with the cell membrane; amylase contents deposited outside of the cell.

4.3 Gaseous exchange in humans

After studying this section you should be able to:

- *understand why organisms need to be adapted for gaseous exchange*
- *explain how the human breathing system brings about ventilation and gaseous exchange*
- *recall where gaseous exchange occurs in a dicotyledonous leaf, the gills of a bony fish and the tracheal system of an insect*
- *show awareness of the adaptations of leaves, gills and tracheoles for efficient gaseous exchange*

LEARNING SUMMARY

How are organisms adapted for efficient gaseous exchange?

AQA	1.4 2.7
CCEA	2.1
EDEXCEL	2.6
OCR	1.2.1
WJEC	2.2

The range of respiratory surfaces in this chapter each have common properties, such as high surface area to volume ratio, one cell thick lining tissue, many capillaries.

The exchange of substances across cell surface membranes has been described. Larger organisms have a major problem in exchange because of their low surface area to volume ratio. Some organisms, like flatworms, have a large surface area due to the shape of their body. Other organisms satisfy their needs by having tissues and organs which have special adaptations for efficient exchange. In simple terms, these structures achieve a very high surface area, e.g. a leaf, and link to the transport system to allow import and export from the organ.

Gaseous exchange in humans

AQA	1.4
CCEA	2.1
EDEXCEL	2.6
OCR	1.2.1
WJEC	2.2

The diagram below shows the **human gas exchange system**. The alveoli are the site of gaseous exchange and they are connected to the outside air via a system of branching tubes.

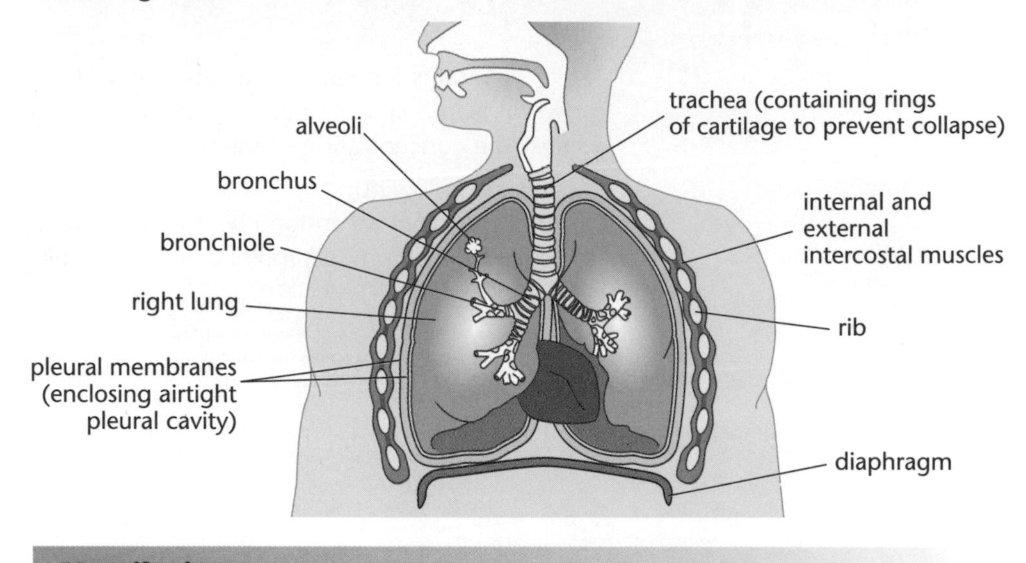

Ventilation

Drawing air in and out of the lungs involves changes in pressure and volume in the chest. These changes work because the **pleural** membranes form an airtight **pleural cavity**.

Breathing in (inhaling):

1. The external intercostal muscles contract, moving the ribs upwards and outwards.
2. The diaphragm contracts and flattens.
3. Both of these actions will increase the volume in the pleural cavity and so decrease the pressure.
4. Air is therefore drawn into the lungs.

Breathing out (exhaling):

1. The internal intercostal muscles relax and the ribs move down and inwards.
2. The diaphragm relaxes and domes upwards.
3. The volume in the pleural cavity is decreased so the pressure is increased.
4. Air is forced out of the lungs.

4.4 Gaseous exchange in other organisms

Lungs of a mammal

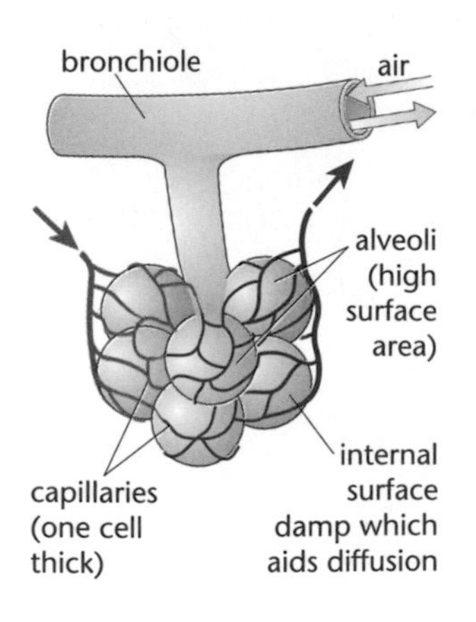

The ventilation mechanism allows inhalation of air, which then diffuses into alveoli to exchange the respiratory gases. Completion of ventilation takes place when gases are expelled into the atmosphere. The diagram on the left shows the structure of the alveoli.

Adaptations of lungs for gaseous exchange

- Air flows through a trachea (windpipe) supported by cartilage.
- It reaches the alveoli via tubes known as bronchi and bronchioles.
- Lungs have many alveoli (air sacs) which have a high surface area.
- Each alveolus is very thin (diffusion is faster over a short distance).
- Each alveolus has an inner film of moisture containing a chemical called surfactant. This reduces the surface tension and makes it easier to inflate the lungs.

- Each alveolus has many capillaries, each one cell thick, to aid diffusion.
- There are many blood vessels in the lungs to give a high surface area for gaseous exchange and transport of respiratory substances.

Measuring ventilation

The **maximum** volume of air that can be breathed out in one breath is called the vital capacity.

The process of ventilation can be investigated using a device called a **spirometer**. It can measure the volume of air exchanged in a single breath. This is called the **tidal volume.** It can also measure the number of breaths per minute – the **breathing rate.** If these two figures are multiplied together, the result gives the volume of gas exchanged in one minute, the **pulmonary ventilation**.

pulmonary ventilation = tidal volume × breathing rate

A dicotyledonous leaf

AQA	2.7
CCEA	1.8
WJEC	2.2

The diagram below shows a section through a leaf. Leaves of plants give a high surface area over which exchange takes place. Specialised tissues increase the efficiency of exchange to allow photosynthesis to supply the plant with enough energy-rich carbohydrates.

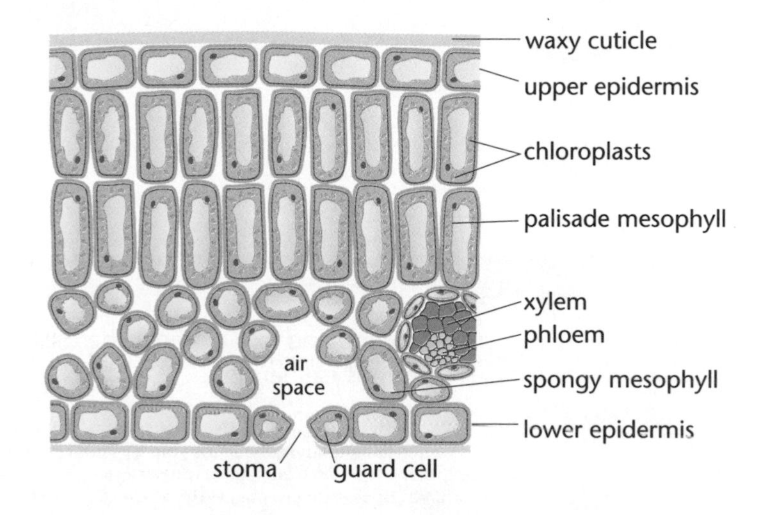

Adaptations of a leaf for photosynthesis

- A flat, thin blade (lamina) to allow maximum light absorption.
- Cells of the upper epidermis have a waxy cuticle to reflect excess light but allow entry of enough light for photosynthesis.
- Each leaf has many chloroplasts to absorb a maximum amount of light.
- Chloroplasts contain many thylakoid membranes, stacked in grana to give a high surface area to absorb the maximum quantity of light.
- Palisade cells, containing chloroplasts, pack closely together to 'capture' the maximum amount of light.
- Many guard cells open stomata to allow carbon dioxide in and oxygen out during photosynthesis.
- Air spaces in the mesophyll store lots of carbon dioxide for photosynthesis or lots of oxygen for respiration.
- Xylem of the vascular bundles brings water to the leaf for photosynthesis.
- Phloem takes the carbohydrate away from the leaf after photosynthesis.

The gills of a bony fish

AQA 2.7
WJEC 2.2

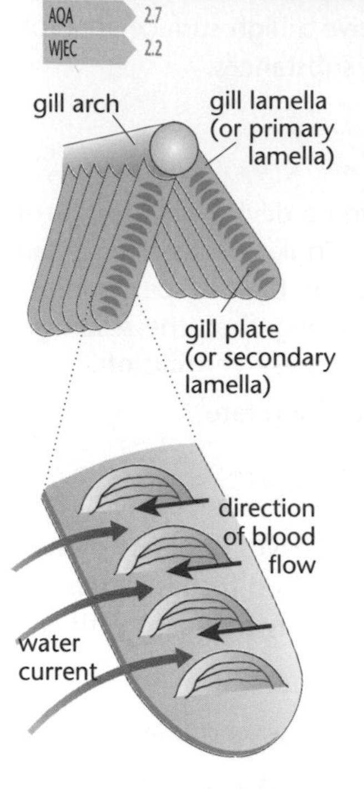

The ventilation mechanism of a fish allows intake of water, and passes it across the gills. The diagram on the left shows the structures of the gills which allow maximum exchange to take place.

Adaptations of gills for gaseous exchange

- The gills of a bony fish have a very high surface area to volume ratio.
- Gills consist of many flat gill filaments, stacked on top of each other, to give a high surface area for maximum exchange.
- Each gill filament has many gill plates which further increase surface area.
- Gill plates are very thin and full of blood capillaries to aid exchange.
- The gradients of O_2 and CO_2 are kept at a maximum by the **counter-current** flow mechanism. By allowing water to flow over the gills in an opposite direction to blood, maximum diffusion rate is achieved.

All respiratory surfaces are damp to allow effective transport across cell membranes.

Progress check

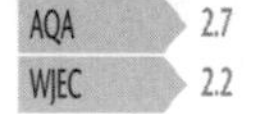

Describe and explain how the gills of a bony fish are adapted for efficient gaseous exchange.

Each gill consists of many thin gill filaments, stacked on top of each other which give a high surface area to volume ratio, for maximum exchange of gases; each gill filament has many gill plates which further increase surface area; each gill plate is very thin and full of blood capillaries; the gradients of O_2 and CO_2 are kept at a maximum by counter-current flow; water flows over the gills in an opposite direction to blood to achieve maximum diffusion rates.

The tracheal system of insects

AQA 2.7
WJEC 2.2

An insect is covered by an exoskeleton which is covered in wax and is impermeable. Insects have a series of holes in the cuticle called spiracles. These spiracles lead into a set of tubes called the tracheal system. The larger tubes, called tracheae, branch into smaller tubes called tracheoles. These tiny tubes are permeable and end on the cells of the body. Oxygen can therefore diffuse from the air into the cells. This system helps reduce water loss from the insect.

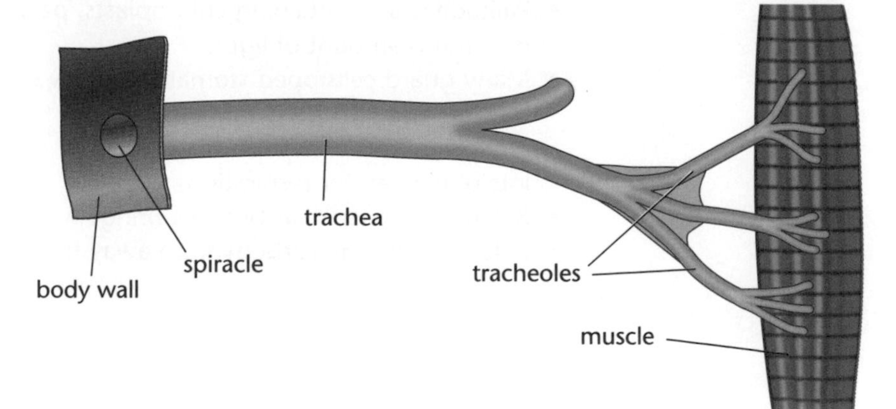

Sample question and model answer

The diagram below shows two adjacent plant cells A and B.
The water potential equation is:

$\psi(\text{cell}) = \psi s + \psi p$

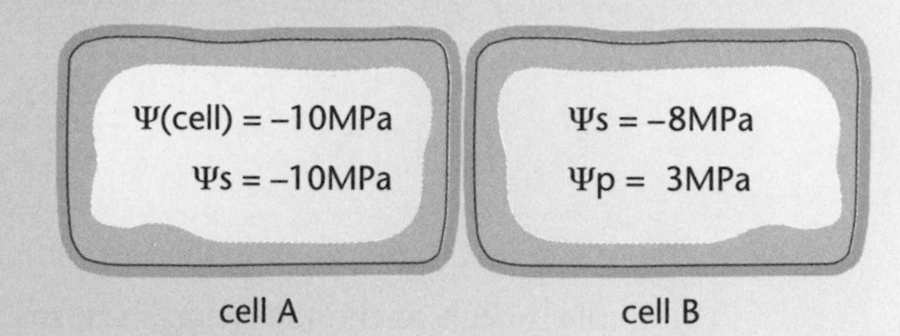

(a) (i) Calculate the water potential of cell B. [1]

Using the equation:
$\psi(\text{cell}) = \psi s + \psi p$
$= -8 \text{ MPa} + 3 \text{ MPa}$
$= -5 \text{ MPa}$

In this question you were given the equation. Try to remember it because it could earn you a mark. Additionally, if there were calculations you need the equation to access the other marks.

When given any two of the values you can work out the other.

(ii) Draw an arrow on the diagram to show the direction of water flow. Show how you worked out the direction. [2]

The arrow should be drawn from cell B to cell A.
Direction from –5 MPa to –10 MPa.

(iii) What is the value of the pressure potential (ψp) of cell A? [1]

$\psi(\text{cell}) = \psi s + \psi p,$
$-10 \text{ MPa} = -10 \text{ MPa} + \psi p$
$\psi p = -10 \text{ MPa} + 10 \text{ MPa}$
$= 0 \text{ MPa}$

Note that from total plasmolysis up to incipient plasmolysis the resistance of the cell, i.e. ψp is zero.

Only when the cell membrane contacts with the wall does it have an effect.

(iv) Name the condition of the cell when $\psi(\text{cell}) = 0$ [1]

full turgor or fully turgid

(b) Give **one** difference between the following terms: [2]

facilitated diffusion

molecules move down a gradient

active transport

energy is needed for the process

Be careful with this type of question. You may believe that 'up a gradient' could be given for active transport. It is correct, but it's too close to the 'down a gradient' idea for facilitated diffusion.

Go for a completely different idea, as shown.

(c) What effect would the following temperatures have on the active transport of Mg^{2+} ions across a cell surface membrane of a plant cell? Assume the plant is a British native. [4]

(i) 30°C

It is likely that active transport would be efficient because the temperature would be ideal for the Mg^{2+} to bind with a carrier molecule.

(ii) 80°C

process likely not to work;
protein carrier denatured;
Mg^{2+} would not be able to bind.

Practice examination questions

1

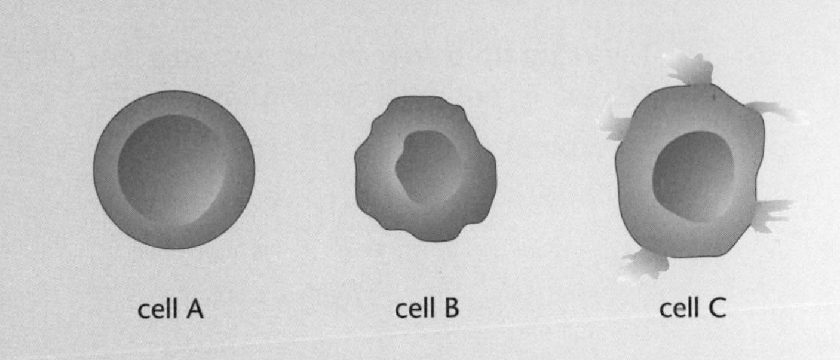

Cells A, B and C have been placed in different concentrations of salt solution.

(a) Explain each of the following in terms of water potential.

(i) Cell A did not change size at all.

(ii) Cell B decreased in volume.

(iii) Cell C became swollen and burst. [3]

(b) Which process is responsible for the changes to cells B and C? [1]

2 (a) Give one similarity between active transport and facilitated diffusion. [1]

(b) Give one difference between active transport and facilitated diffusion. [1]

3 The diagram shows a section through a leaf.

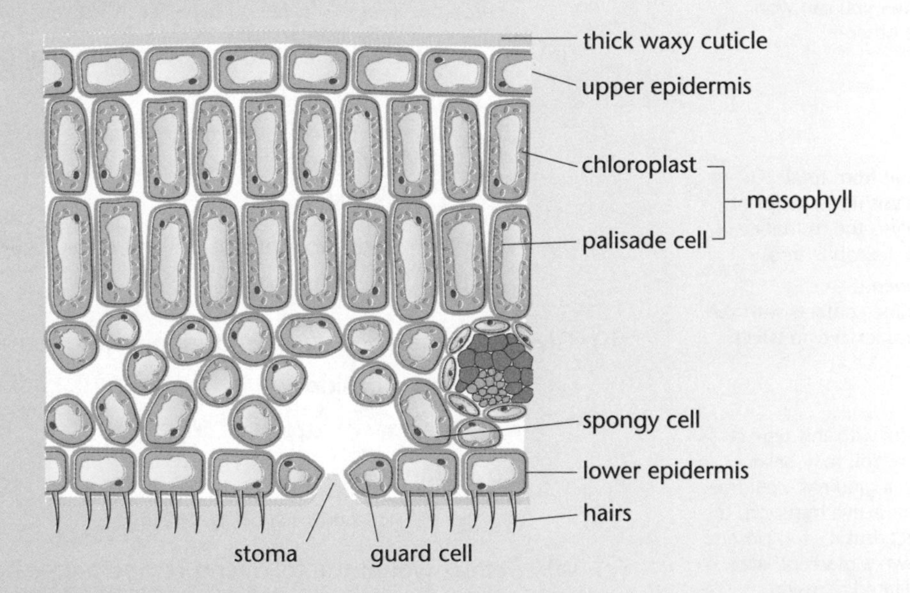

Describe how the leaf is adapted for efficient photosynthesis. [6]

4 Describe and explain how the following are adapted to efficient gaseous exchange.

(a) The alveoli of lungs. [4]

(b) The gills of a fish. [4]

Chapter 5

Transport

The following topics are covered in this chapter:

- *Mass transport systems*
- *Heart: structure and function*
- *Blood vessels*
- *The transport of substances in the blood*
- *The transport of substances in a plant*

5.1 Mass transport systems

After studying this section you should be able to:

- *explain why most multicellular organisms need a mass transport system*
- *understand the importance of a high surface area to volume ratio*
- *describe the differences between open and closed circulatory systems*
- *understand the implications of using single or double circulatory systems*

LEARNING SUMMARY

Why do most multicellular organisms need a mass transport system?

AQA	2.7
CCEA	2.1
EDEXCEL	1.3.6
OCR	1.2.2-3
WJEC	2.3

The bigger an organism is, the lower its surface area to volume ratio. Substances needed by a large organism could not be supplied through its exposed external surface. Oxygen passing through an external surface would be rapidly used up before reaching the many layers of underlying cells. Similarly waste substances would not be excreted quickly enough. This problem has been solved, through evolution, by specially adapted tissues and organs.

KEY POINT

Leaves, roots, gills and lungs all have high surface area to volume properties so that supplies of substances vital to **all** the living cells are made available by these structures. Movement of substances to and from these structures is carried out by efficient **mass transport systems.**

This is a little like transport on trains where people travel together on the same train, in the same direction, at the same speed, but may get off at different places.

In a mass transport system, all the substances move in the same direction at the same speed. Across the range of multicellular organisms found in the living world are a number of mass transport systems, e.g. the mammalian circulatory system and the vascular system of a plant.

Mass transport systems are just as important for the rapid removal of waste as they are for supplies. Supplies include an immense number of substances, e.g. glucose, oxygen and ions. Even communication from one cell to another can take place via a mass transport system, e.g. hormones in a blood stream.

The greater the metabolic rate of an organism, the greater the demands on its mass transport system.

Different types of circulatory systems

OCR	1.2.2
WJEC	2.3

Most organisms that are beyond a certain size have a circulatory system. These systems may be open or closed.

In a **closed circulation** the blood is contained in blood vessels as it circulates. In an **open circulation** the blood is contained in the body cavity.

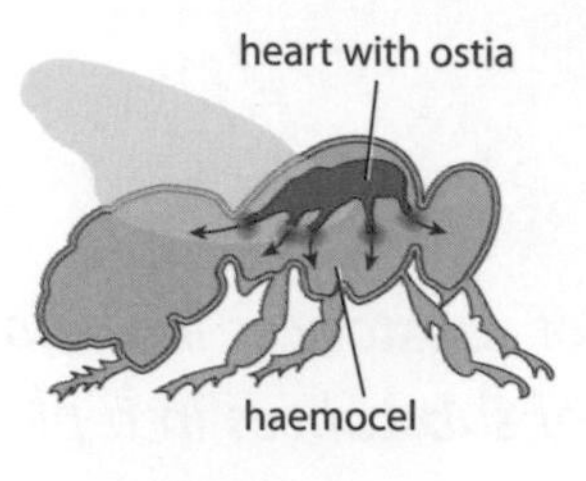

Insects have an open circulation. The blood is in the body cavity called the haemocoel. It does not circulate in blood vessels but a dorsal tube-shaped heart maintains movement of the blood in the chamber.

earthworm
dorsal vessel
hearts
capillary system
ventral vessel

Earthworms have a closed circulation. Five of the blood vessels act like hearts, pumping the blood through the main two blood vessels.

In vertebrates, the pumping of the blood is performed by a specialised heart.

KEY POINT

Vertebrate circulatory systems can be either single or double and this results in structural differences between their hearts.

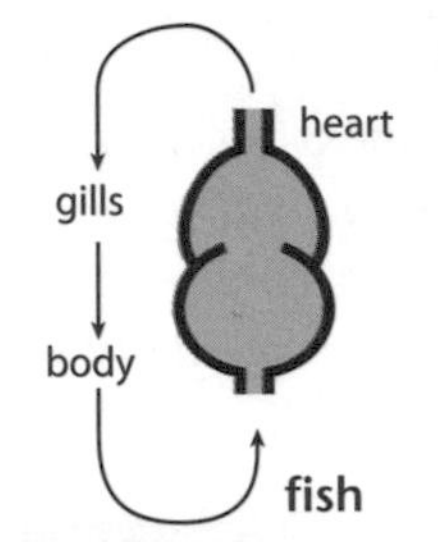

In a single circulation, the blood has to pass through two capillary beds, one after the other. This makes blood flow more slowly compared with a double circulation.

Fish have a **single circulatory system.** This means that the blood flows through the heart once on each circuit of the body.

Mammals have a **double circulatory** system. This means that as blood enters the heart it is pumped to the lungs, exchanges carbon dioxide for oxygen, and returns to the heart where further pumping propels it through the rest of the body. The blood moves through the heart twice during each cardiac cycle.

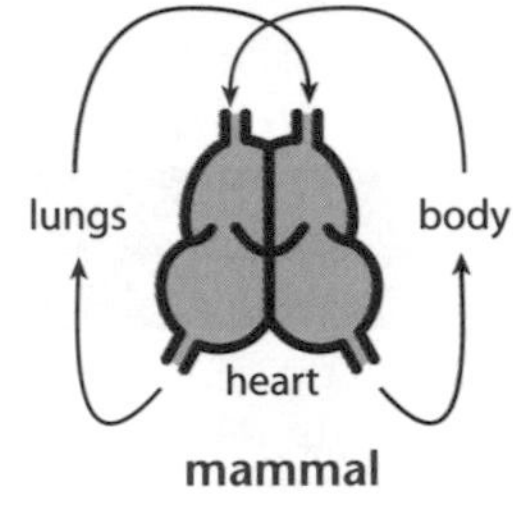

5.2 Heart: structure and function

LEARNING SUMMARY

After studying this section you should be able to:

- *recall the structure, cardiac cycle and electrical stimulation of a mammalian heart*

The mammalian heart

AQA	1.5
CCEA	2.1
EDEXCEL	1.3.7
OCR	1.2.2
WJEC	2.3

The heart consists of a range of tissues. The most important one is cardiac muscle. The cells have the ability to contract and relax through the complete life of the person, without ever becoming fatigued. Each cardiac muscle cell is **myogenic.** This means it has its own inherent rhythm. Below are diagrams of the heart and its position in the circulatory system.

Note that tricuspid and bicuspid valves are known as atrioventricular valves.

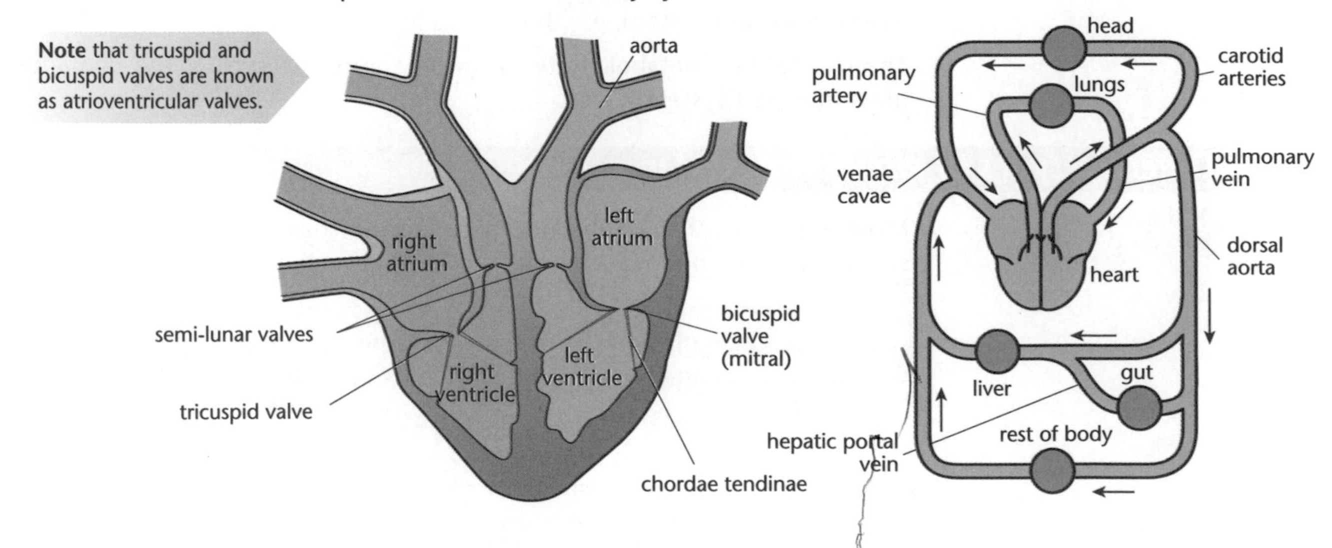

Structure

If blood moved in the wrong direction, then transport of important substances would be impeded.

The heart consists of four chambers, **right** and **left atria** above **right** and **left ventricles**. The functions of each part are as follows.

- The **right atrium** links to the **right ventricle** by the **tricuspid valve**. This valve prevents backflow of the blood into the atrium above, when the ventricle contracts.
- The **left atrium** links to the **left ventricle** by the **bicuspid valve (mitral valve)**. This also prevents backflow of the blood into the atrium above.
- The **chordae tendonae** attach each ventricle to its **atrioventricular valve**. Contractions of the ventricles have a tendency to force these valves up into the atria. Backflow of blood would be dangerous, so the chordae tendonae hold each valve firmly to prevent this from occurring.
- Semi-lunar (pocket) valves are found in the blood vessels leaving the heart (pulmonary artery and aorta). They only allow exit of blood from the heart through these vessels following ventricular contractions. Elastic recoil of these arteries and relaxation of the ventricles closes each semi-lunar valve.
- Ventricles have thicker muscular walls than atria. When each atrium contracts it only needs to propel the blood a short distance into each ventricle.
- The left ventricle has even thicker muscular walls than the right ventricle. The left ventricle needs a more powerful contraction to propel blood to the systemic circulation (all of the body apart from the lungs). The right ventricle propels blood to the nearby lungs. The contraction does not need to be so powerful.

Check out these diagrams of a valve.

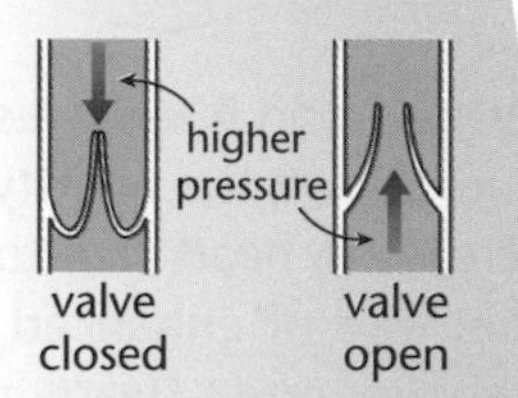

You can work out if a valve is open or closed in terms of pressure. Higher pressure above than below a semi-lunar valve closes it. Higher pressure below the semi-lunar valve than above, opens it.

Cardiac cycle

Blood must continuously be moved around the body, collecting and supplying vital substances to cells as well as removing waste from them. The heart acts as a pump using a combination of **systole** (contractions) and **diastole** (relaxation) of the chambers. The cycle takes place in the following sequence.

Examination questions often test your knowledge of the opening and closing of valves. Always analyse the different pressures given in the question. A greater pressure behind a valve opens it. A greater pressure in front closes it.

Stage 1

Ventricular diastole, atrial systole

Both ventricles relax simultaneously. This results in lower pressure in each ventricle compared to each atrium above. The atrioventricular valves open partially. This is followed by the atria contracting which forces blood through the atrioventricular valves. It also closes the valves in the vena cava and pulmonary vein. This prevents backflow of blood.

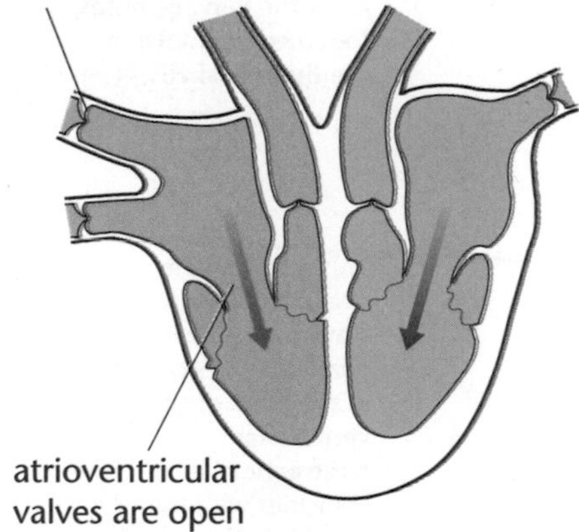

Stage 2

Ventricular systole, atrial diastole

Both atria then relax. Both ventricles contract simultaneously. This results in higher pressure in the ventricles compared to the atria above. The difference in pressure closes each atrioventricular valve. This prevents backflow of blood into each atrium. Higher pressure in the ventricles compared to the aorta and pulmonary artery opens the semi-lunar valves and blood is ejected into these arteries. So blood flows through the systemic circulatory system via the aorta and vena cava and through the lungs via the pulmonary vessels.

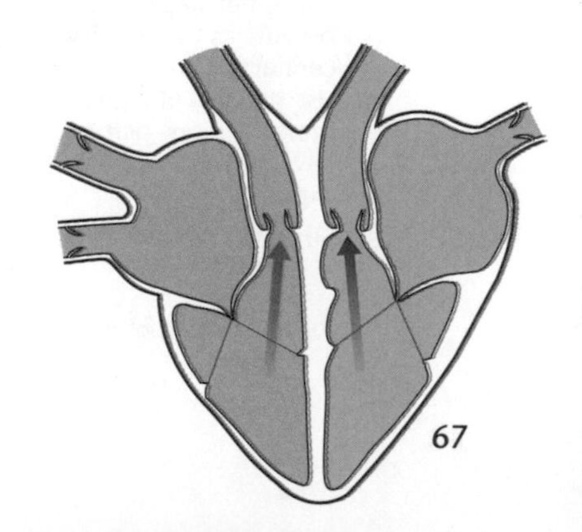

Stage 3

Ventricular diastole, atrial diastole
Immediately following ventricular systole, both ventricles and atria relax for a short time. Higher pressure in the aorta and pulmonary artery than in the ventricles closes the semi-lunar valves. This prevents the backflow of blood. Higher pressure in the vena cava and pulmonary vein than in the atria results in the refilling of the atria.

The cycle is now complete – *GO BACK TO STAGE 1!*

Returning to Stage 1, the cycle begins again. The hormone adrenaline increases the heart rate still further. Even your examinations may increase your heart rate!

The whole sequence above is **one** cardiac cycle or heartbeat and it takes less than one second. The number of heartbeats per minute varies to suit the activity of an organism. Vigorous exercise is accompanied by an increase in heart rate to allow faster collection, supply and removal of substances because of enhanced blood flow. Conversely during sleep, at minimum metabolic rate, heart rate is correspondingly low because of minimum requirements by the cells.

How is the heart rate controlled?

SAN

AVN

Purkinje tissue

The SAN is the natural pacemaker of the heart.

It has already been stated that the cardiac muscle cells have their own inherent rhythm. Even an individual cardiac muscle cell will contract and relax on a microscope slide under suitable conditions. An orchestra would not be able to play music in a coordinated way without a conductor. The cardiac muscle cells must be similarly coordinated, by a **pacemaker** area in the heart. Electrical stimulation from the brain can alter the activity of the pacemaker and therefore change the rate and strength of the heartbeat.

- The heart control centre in the brain is in the medulla oblongata.
- The sympathetic nerve stimulates an increase in heart rate.
- The vagus nerve stimulates a decrease in heart rate.
- These nerves link to a pacemaker structure in the wall of the right atrium, the **sinoatrial node (SAN)**.
- A wave of electrical excitation moves across both atria.
- They respond by contracting (the right one slightly before the left).
- The wave of electrical activity reaches a second pacemaker, the **atrioventricular node (AVN)**, which conducts the electrical activity through the **Purkinje fibres**.
- These Purkinje fibres pass through the septum of the heart deep into the walls of the left and right ventricles.
- The ventricle walls begin to contract from the apex (base) upwards.
- This ensures that blood is ejected efficiently from the ventricles.

All of the Purkinje fibres together are known as the **Bundle of His**.

This is one of the examiners' favourite ways to test heart-related concepts. Look at the **peak** of the **ventricular contraction.** It coincides with the **trough** in the **ventricular volume.** This is not surprising, because as the ventricle contracts it empties! Use the data of higher pressure in one part and lower in another to explain:

(a) movement of blood from one area to another
(b) the closing of valves.

Graphs to show the changes in pressure and volume during the cardiac cycle

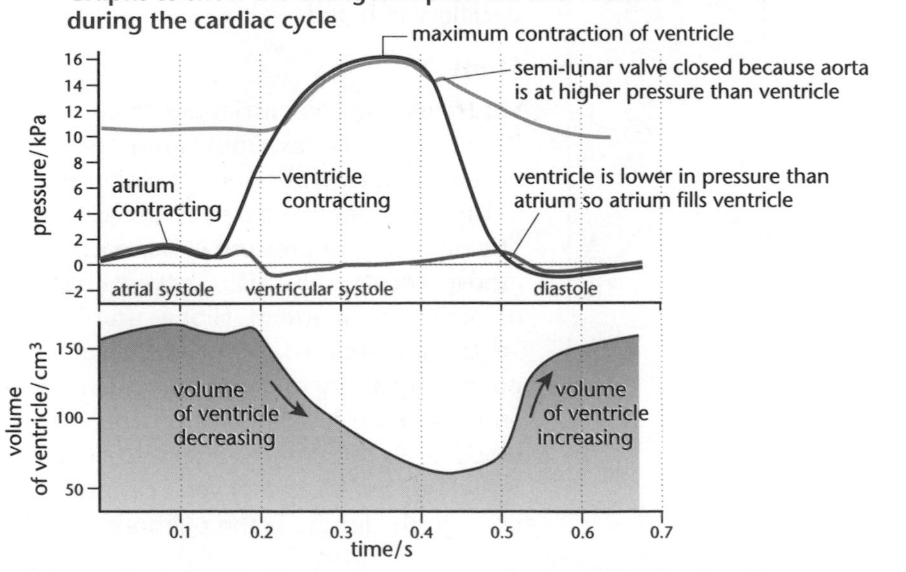

Progress check

The medulla oblongata can increase the heart rate. The statements below include all of the events which take place, but in the wrong order. Write them out in the correct sequence.

A this ensures that blood is ejected efficiently from the ventricles

B the wave of electrical activity reaches the **atrioventricular node (AVN)** which conducts the electrical activity through the **Purkinje fibres**

C a wave of electrical excitation moves across both atria

D the sympathetic nerve conducts electrical impulses

E electrical impulses are received at the **sinoatrial node (SAN)**

F as a result the atria contract

G the ventricle walls begin to contract from the apex (base) upwards

D E C F B G A

5.3 Blood vessels

After studying this section you should be able to:

- *describe the structure and functions of arteries, capillaries and veins*
- *understand the importance of valves in the return of blood to the heart*
- *understand the difference between plasma, tissue fluid and lymph*

LEARNING SUMMARY

Arteries, veins and capillaries

AQA	2.7
EDEXCEL	1.3.8
OCR	1.2.2
WJEC	2.3
CCEA	2.1

The blood is transported to the tissues via the vessels. The main propulsion is by the ventricular contractions. Blood leaves the heart via the arteries, reaches the tissues via the capillaries, then returns to the heart via the veins. Each blood vessel has a space through which the blood passes; this is the **lumen**. The structure of the vessels is shown below.

Artery

- It has a thick **tunica externa** which is an outer covering of tough collagen fibres.
- It has a **tunica media** which is a middle layer of **smooth muscle** and **elastic fibres.**
- It has a lining of **squamous endothelium** (very thin cells).
- It can contract using its **thick muscular layer.**

tunica externa (collagen fibres)
lumen
tunica media
endothelial lining

Note that the pressure in the **arteries** is highest because:

(a) they are closest to the ventricles
(b) they contract forcefully themselves.

Capillaries are the next highest in pressure, the main factor being their resistance to blood flow.

Finally, the pressure of **veins** is the lowest because:

(a) they are furthest from the ventricles
(b) they have a low amount of muscle.

If given blood pressures of vessels, be ready to predict the correct direction of blood flow.

Capillary

- It is a very thin blood vessel, the endothelium is just one cell thick.
- Substances can exchange easily.
- It has such a high resistance to blood flow that blood is slowed down. This gives more time for efficient exchange of chemicals at the tissues.

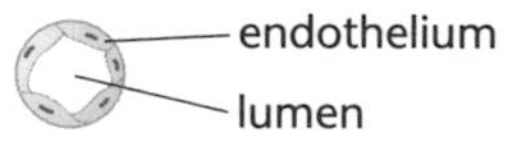

Vein

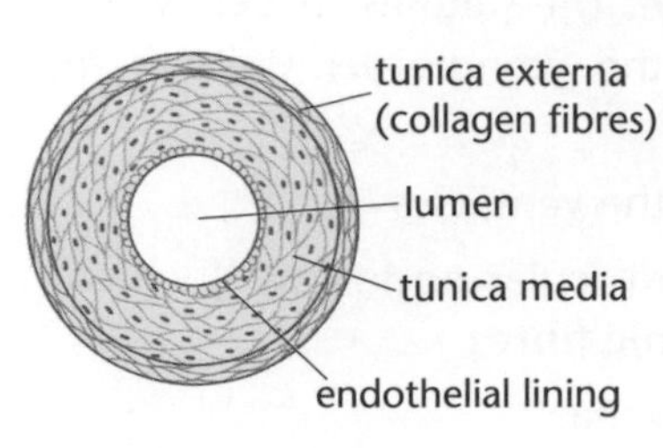

- It has a thin **tunica externa** which is an outer covering of tough collagen fibres.
- It has a very thin **tunica media** which is a middle layer of **smooth muscle** and **elastic fibres.**
- It has a lining of **squamous endothelium** (very thin cells).
- It is lined with semi-lunar valves which prevent the backflow of blood.

How do the veins return the blood to the heart?

direction of blood flow

semi-lunar valve

Veins have a thin tunica media, so only mild contractions are possible. They return blood in an unexpected way. Every time the organism moves physically, blood is squeezed between skeletal muscles and forced along the vein.

> **KEY POINT**
> Blood must travel towards the heart because of the direction of the semi-lunar valves. Any attempt at backflow and the semi-lunar valves shut tightly!

Capillary network

Every living cell needs to be close to a **capillary**. The arteries transport blood from the heart but before entry into the capillaries it needs to pass through smaller vessels called **arterioles.** Many arterioles contain a ring of muscle known as a **pre-capillary sphincter**. When this is contracted the constriction shuts off blood flow to the capillaries, but when it is dilated, blood passes through. Some capillary networks have a shunt vessel. When the sphincter is constricted blood is diverted along the shunt vessel so the capillary network is by-passed. After the capillary network has permeated an organ the capillaries link into a **venule** which joins a **vein**.

In the skin the superficial capillaries have the arteriole/shunt vessel/venule arrangement as shown opposite. When the arteriole is dilated (**vasodilation**) more heat can be lost from the skin. When the arteriole is constricted (**vasoconstriction**) the blood cannot enter the capillary network so is diverted to the core of the body. Less heat is lost from the skin.

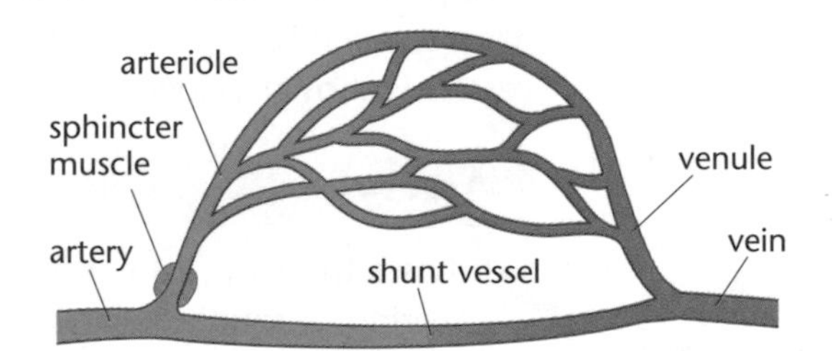

Although the pressure of the blood in the capillaries is lower than in the arteries or arterioles, there is still enough pressure to force out some of the liquid part of the blood. The liquid part of the blood is called **plasma** and when it is forced out of the capillaries it is called **tissue fluid**.

This tissue fluid bathes the cells, supplying them with nutrients and taking up waste products. At the venous end of the capillary bed, most of this tissue fluid is reabsorbed back into the capillaries.

Lymphatic system

There is a network of vessels other than the blood system. They are the **lymphatic vessels**. They collect any tissue fluid that is not reabsorbed back into the capillaries.

The lymph vessels have valves to ensure transport is in one direction. Along some parts of the lymphatic system are lymph nodes. These are swellings lined with white blood cells (macrophages and lymphocyte cells). The lymph fluid is finally emptied back into the blood near the heart.

Progress check

The diagram shows the structure of a blood vessel.

(a) (i) Which type of vessel, artery, vein or capillary is shown? Give a reason for your choice.

(ii) What is the function of the tunica media?

(b) The pressure values 30 kPa, 10 kPa and 5 kPa correspond to the different types of vessel. Give the correct value for each vessel so that blood flows around the body.

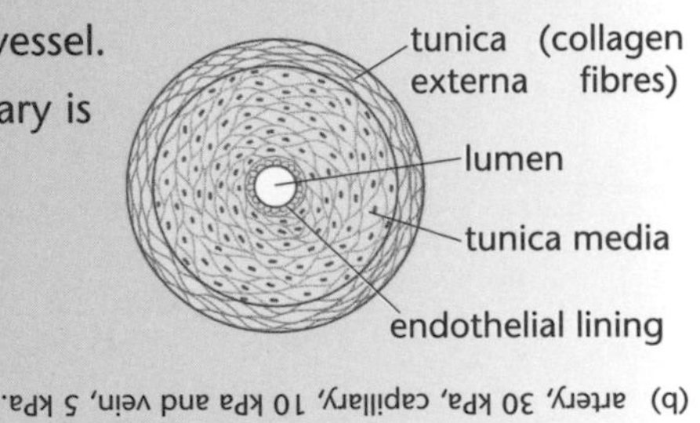

(a) (i) artery; the vessel has a thick tunica externa
(ii) contracts to help transport blood.
(b) artery, 30 kPa, capillary, 10 kPa and vein, 5 kPa.

5.4 The transport of substances in the blood

After studying this section you should be able to:

LEARNING SUMMARY

- *describe the transport of oxygen in the blood and explain how oxygen is released at the tissues*
- *describe the transport of carbon dioxide*

How is oxygen transported?

AQA 2.4
CCEA 2.1
OCR 1.2.2
WJEC 2.3

Oxygen is absorbed in the lungs from fresh air which has been breathed in. Red blood cells **(erythrocytes)** contain the protein **haemoglobin** which can reversibly combine with oxygen. In the lungs, where the concentration of oxygen is high, haemoglobin will take up oxygen and form oxyhaemoglobin. In the tissues where the oxygen concentration is low, the oxyhaemoglobin will dissociate and release the oxygen. This is shown by the graph below which is called the oxygen dissociation curve.

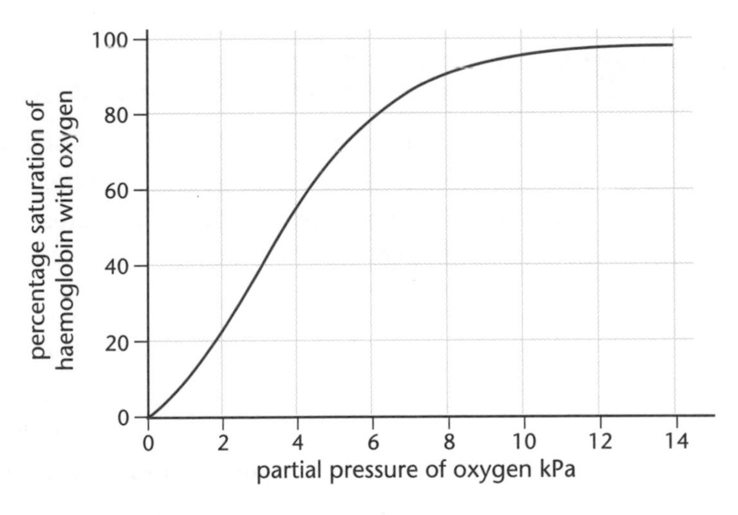

Features of the dissociation curve

- At high partial pressures of oxygen, haemoglobin has a high affinity (attraction) for oxygen and is highly saturated.
- At low partial pressures, the affinity is lower and the oxyhaemoglobin dissociates and is less saturated.
- The curve is sigmoid or 'S' shaped. This means that the curve is steep and a small change in partial pressure causes a massive loading or unloading of oxygen.

Changes to the dissociation curve

Different factors can cause changes to the dissociation curve:

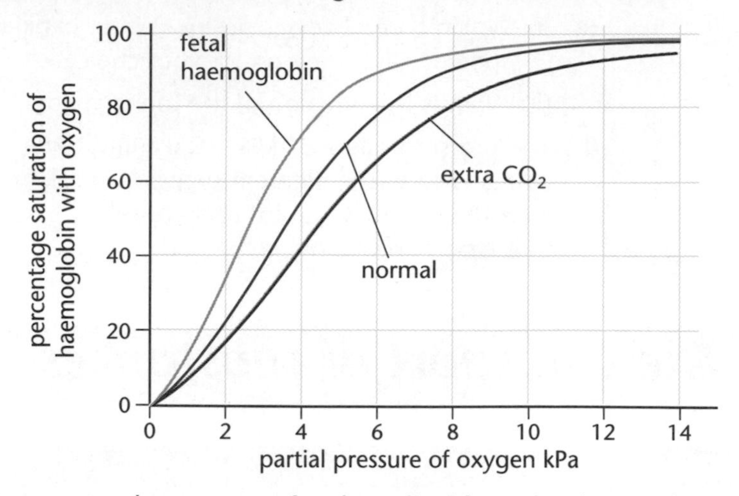

- The greater the amount of carbon dioxide at the tissues, the more the dissociation curve is moved to the right, and the more oxygen is 'off-loaded' to the tissues. This is called the Bohr shift. The carbon dioxide lowers the affinity of the haemoglobin for oxygen.
- Fetal haemoglobin has a greater affinity for oxygen than adult haemoglobin. This allows the fetus to take oxygen from the mother's haemoglobin.

Myoglobin only has one polypeptide chain compared to four found in each molecule of haemoglobin. Myoglobin is often found in diving animals or animals that live in anaerobic mud, e.g. lugworms.

Myoglobin

There is another substance that can act as a respiratory pigment. This is **myoglobin.** It does not travel in the blood but is found in muscle. It has a greater affinity for oxygen than haemoglobin and so only releases oxygen at very low partial pressures. It acts as a store of oxygen trying to prevent anaerobic respiration from occurring.

Transport of carbon dioxide

AQA	2.4
OCR	1.2.2
WJEC	2.3
CCEA	2.1

This is done with the help of the red blood cells as follows:

- carbon dioxide diffuses into red blood cells from the tissues
- the carbon dioxide reacts with water to produce carbonic acid, this reaction being catalysed by the enzyme **carbonic anhydrase** in the cell *(a very fast reaction!)*

carbonic anhydrase

$$H_2O + CO_2 \rightleftharpoons H_2CO_3$$

water carbon dioxide ⇌ carbonic acid

- the carbonic acid **ionises** into H^+ and HCO_3^-

$$H_2CO_3 \rightleftharpoons H^+ + HCO_3^-$$

- haemoglobin combines with H^+ ions forming **haemoglobinic acid** which is very weak

$$H^+ + Hb \rightleftharpoons HHb$$

- HCO_3^- ions diffuse into the blood plasma to be transported to the lungs
- Cl^- ions diffuse into the red blood cell from the plasma; this counteracts the build up of positive charge from the H^+ ions. This is known as the **chloride shift.**

The whole process is reversed once the blood reaches the lungs.

Most of the carbon dioxide in the blood is carried in this way as HCO_3^- ions. However, a small amount is carried combined with haemoglobin as carbaminohaemoglobin. Some is also dissolved in the plasma.

Do not confuse carbaminohaemoglobin with carboxyhaemoglobin which is formed when carbon monoxide combines with Hb.

Plasma

AQA	2.4
OCR	1.2.2
WJEC	2.3
CCEA	2.1

This is the fluid in which all of the blood contents are transported. Listed below are some substances transported in the plasma:

- **water** – dissolves substances such as glucose for transport, stores dissolved prothrombin and fibrinogen which may be used later in clotting
- **proteins** – some are used to buffer the pH of the blood
- **glucose** – on its way to releasing energy in respiration
- **lipids** – on their way to releasing energy in respiration
- **amino acids** – on their way to cells to help assemble proteins or release energy in respiration
- **salts** – contribute to the water potential of blood, so that cells are not dehydrated by osmosis
- **hormones** – chemical messenger-molecules on their way to a target organ
- **antigens** – recognition proteins preventing white blood cells from destroying the person's own blood
- **antibodies** – made by lymphocytes to destroy antigens
- **urea** – made in the liver from excess amino acids, extracted by the kidneys.

Blood has a major role in the defence against disease (see the immune system page 111).

Water has many important functions in the body, including being transported to the sweat glands to cool the body down.

Note that the list outlines just some of the functions of plasma-transported substances. There are many more!

5.5 The transport of substances in a plant

After studying this section you should be able to:

- *recall the structure of a root and understand how water and ions are absorbed*
- *recall the structure of xylem and phloem and explain the processes by which they transport essential chemicals*
- *understand how plants lose water and how this loss can be measured*
- *recall the adaptations of xerophytes*

LEARNING SUMMARY

Root structure and functions

AQA	2.7
CCEA	1.8 / 2.1
OCR	1.2.3
WJEC	2.3

The roots of a green plant need to exchange substances with the soil environment. The piliferous zone just behind a root tip has many root hairs which have a high surface area to volume ratio.

- Root hairs are used for absorption of water and mineral ions and the excretion of carbon dioxide.
- They have a cell membrane with a high surface area to volume ratio to efficiently absorb water, mineral ions and oxygen, and excrete carbon dioxide.
- They project out into the soil particles which are surrounded by soil water at high water potential compared with the low water potential of the contents of the root hairs.
- They have a cell membrane which is partially permeable to allow water absorption by osmosis (see page 57).
- As they absorb more water by osmosis, the cell sap becomes more dilute compared with neighbouring cells. Water therefore moves to these adjacent cells which become more diluted themselves, so osmosis continues across the cortex.
- They have carrier proteins in the cell membranes to allow mineral ions to be absorbed by active transport.

Note that the root hairs also absorb oxygen from the air to aid aerobic respiration. The high surface area to volume ratio certainly helps!

Remember that water moves from a higher water potential to a more negative water potential.

Remember that active transport needs energy, so mitochondria will be close to the carrier molecules on the membranes.

Passage of water into the vascular system

Once absorbed by osmosis, water needs to pass to the xylem vessels in order to move up the plant. First it must move across the cortex of the root and through the endodermis before entering the xylem. The mechanism of passage is not known but there are three theories:

Note the different theories for water transport across the width of the root.

- **apoplast** route, where the water is considered to pass between the cells
- **symplast** route, where the water is considered to pass via the cytoplasm of the cells via **plasmodesmata** (cytoplasmic strands connecting one cell to another)
- **vacuolar** route, where the water is considered to pass through the tonoplast then through the sap vacuole of each cell.

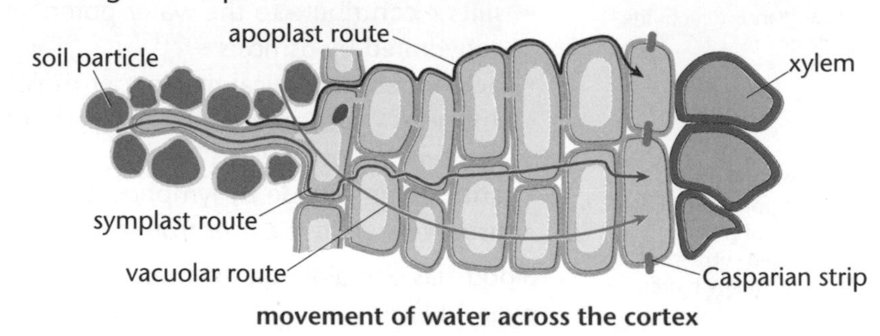

movement of water across the cortex

Casparian strip

Water moves across the cortex and needs to pass through the endodermal cells before entering the xylem vessels of the vascular system. Each cell of the endodermis has a waterproof band around it, just like a ribbon around a box. This means that water must pass through the cell in some way, rather than around the outside. If water moves by the apoplast route up to this point, then it must now move into the symplast or vacuolar pathways.

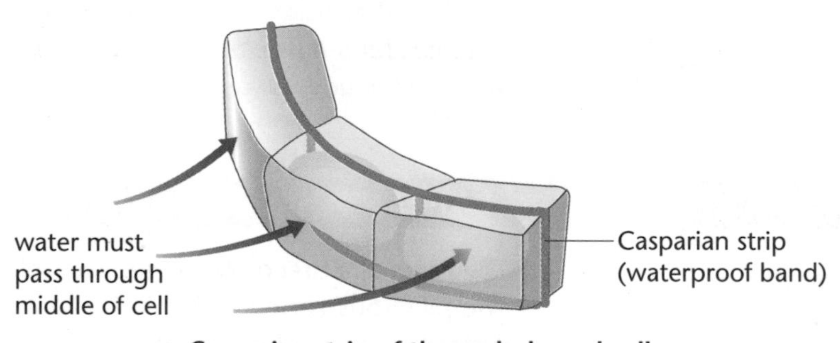

Casparian strip of the endodermal cells

Once the water has passed through the endodermis and navigated the pericycle then it must pass into the xylem for upward movement to the leaves and to the tissues.

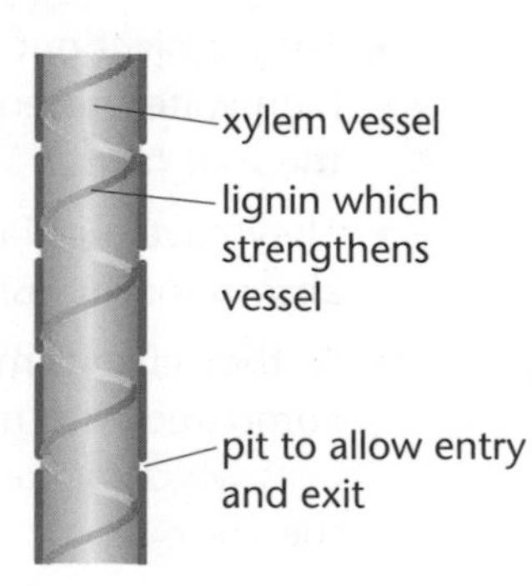

How does water move up the vascular system to the leaves?

Remember that the xylem is part of the mass flow system ensuring that all cells receive their requirements.

Water moves into the **xylem** vessels in the vascular system in the centre of the root; it enters via **bordered pits**. The xylem is internally lined with **lignin**. This substance is waterproof and it also gives great strength to the xylem vessels, which are tube shaped. Much of the strength of a plant comes from cells toughened by lignin. A Giant Redwood tree is many metres high but water is still able to reach all the cells. Water moves up the xylem for the following reasons.

- **Root pressure** gives an initial upward force to water in the xylem vessels. This can be shown by cutting off a shoot near soil level. Some sap will pour vertically out of the xylem of the remaining exposed xylem.
- Water moves up the xylem by **capillarity** which is the upward movement of a fluid in a narrow bore tube – xylem has very narrow vessels.
- Capillarity occurs because the water molecules have an attraction for each other (cohesion) so when one water molecule moves others move with it.
- Capillarity has another component – the fact that the water molecules are attracted to the sides of the vessels pulls the water upwards (adhesion).
- Transpiration causes a very negative water potential in the mesophyll of the leaves. Water in the xylem is of higher water potential and so moves up the xylem.

The factors in the list are known as the cohesion-tension theory and explain how water moves up the xylem.

Xylem vessels die at the end of their maturation phase. The lignin produced inside the cells finally results in death. The young xylem cells end to end, finally produce a long tube-like structure (vessel) through which water passes. Xylem can still transport water after the death of the plant.

Mineral ions are also transported in the xylem.

Translocation

This is an **active process** by which **sugars** and **amino acids** are transported through the phloem. Sugar is produced in the photosynthetic tissues and must be exported from these **sources** to areas of need, i.e. usually areas which have large energy requirements. These areas are called sinks, e.g. terminal buds and roots.

KEY POINT

Roots cannot photosynthesise so they need carbohydrates to be supplied by other parts of the plant such as the leaf or storage organs.

The sugars are transported in the phloem which consists of two types of cell, the **sieve tube** and **companion cell**. Unlike xylem, the cells of the phloem are living.

Structure of the phloem tissue

The sieve tube has no nucleus so that essential proteins for life are made by the companion cell which does possess a nucleus. The companion cell maintains services to the sieve tube.

- Each **sieve tube** links to the next via a **sieve plate** which is perforated with pores.
- The sieve tube has cytoplasm and a few small mitochondria.
- Sugars are thought to pass through the sieve tubes by **cytoplasmic streaming**.
- The sieve tubes have no nucleus but are alive because of **cytoplasmic connections (plasmodesmata)** with the companion cell.
- Each companion cell has a nucleus and mitochondria.

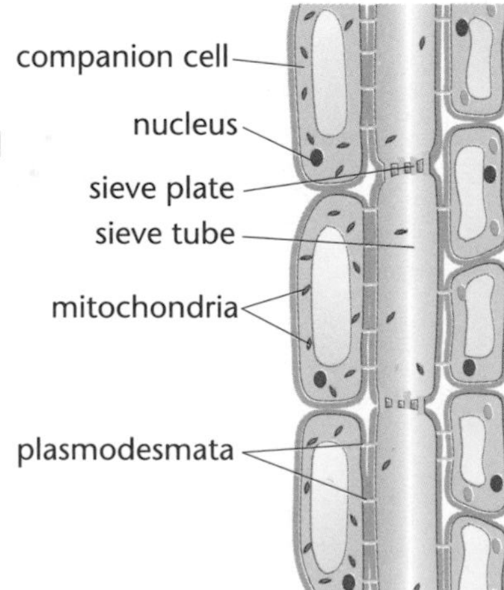

Radioactive labelling

This technique has been used to investigate the mechanism of translocation. It involves the use of a substance such as radioactive CO_2. The radioactive isotope, ^{14}C is used to make $^{14}CO_2$. A leaf is allowed to photosynthesise in the presence of $^{14}CO_2$ and makes the radioactive sugar ($^{14}C_6H_{12}O_6$). The route of the radioactive sugar can be traced using a Geiger-Müller counter. The greater the number of radioactive disintegrations per unit time, the greater the concentration of the sugar in that part of the plant.

How is water lost from a leaf?

AQA	2.7
CCEA	1.8 / 2.1
OCR	1.2.3
WJEC	2.3

Water moves up the xylem and into the mesophyll of a leaf. The process by which water is lost from any region of a plant is **transpiration**. Water can be lost from areas such as a stem, but most water is lost by **evaporation** through the **stomata**. Each stoma is a pore which can be open or closed and is bordered at either side by a guard cell. The diagrams show an open stoma and a closed stoma.

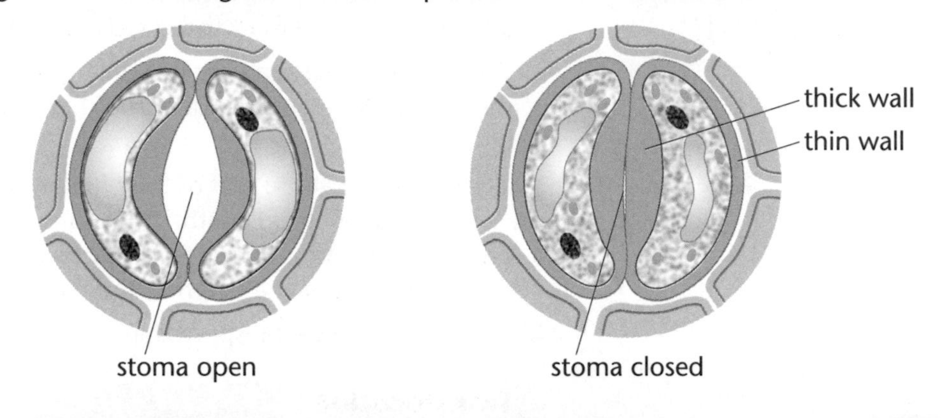

> **KEY POINT**
>
> Transpiration from a leaf takes place as follows:
> - the air spaces in the mesophyll become **saturated** with water vapour **(higher water potential)**
> - the air outside the leaf may be of **lower humidity (more negative water potential)**
> - this causes water molecules to **diffuse** from the mesophyll of the leaf to the outside.

Some water can escape through the cell junctions and membranes. This is known as **cuticular transpiration**. In the dark all stomata are closed. Even so, there is still water loss by cuticular transpiration.

How do the guard cells open and close?

In the presence of light:

- K^+ ions are actively transported into the guard cells from adjacent cells
- malate is produced from starch
- K^+ ions and malate accumulate in the guard cells
- this causes an influx of water molecules
- the cell wall of each guard cell is thin in one part and thick in another
- the increase in hydrostatic pressure leads to the opening of the stomata.

Closing of the stomata is the reverse of this process. Under different conditions the stomata can be partially open. The rate of transpiration can increase in warm, dry conditions or decrease at the opposite extreme.

Measuring the rate of transpiration

This is done indirectly by using a potometer. This instrument works on the following principle: for every molecule of water lost by transpiration, one is taken up by the shoot.

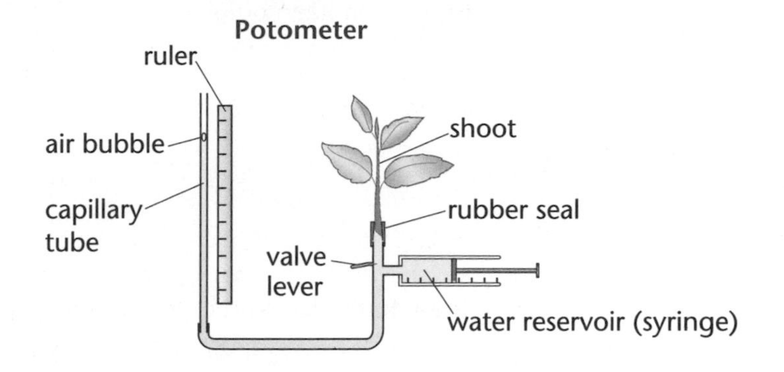

Note that a living shoot may be photosynthesising whilst attached to the instrument. Only a minute amount of water would be used in this process. The instrument gives an accurate measure of transpiration.

The potometer is used as follows:

- a shoot is cut and the end is quickly put in water to prevent an air lock in the xylem
- the potometer is filled under water so that the capillary tube is full
- all air bubbles are removed from the water
- the shoot is put into the rubber seal
- the valve is changed to allow water uptake
- the amount of water taken up by the shoot per unit time is measured
- the shoot can be tested under various conditions.

Xerophytes

AQA	2.7
CCEA	2.1
OCR	1.2.3
WJEC	2.3

These are plants which have **special adaptations** to survive in drying, environmental conditions where many plants would become desiccated and die. The plants survive well because of a combination of the following features:

- thick cuticle to reduce evaporation
- reduced number of stomata
- smaller and fewer leaves to reduce surface area
- hairs on plant to reduce air turbulence
- protected stomata to prevent wind access
- aerodynamic shape to prevent full force of wind
- deep root network to absorb maximum water
- some store of water in modified structures, e.g. the stem of a cactus.

In an exam you may be given a photomicrograph of a xerophytic plant which you have not seen before. Look for *some* of the features covered in the bullet points opposite.

A cactus is a good example of a xerophyte. It makes excellent use of what little water there is available, and holds on to what it does manage to absorb really well. Its cuticle and epidermis are so thick that metabolic water released from the cells during night time respiration is retained for photosynthesis during the day. Nothing is wasted!

Progress check

Water is absorbed into a plant by the root hairs.

(a) The water potential of the root hair cells is more negative than in the soil water. Is this statement true or false?

(b) Describe:

(i) the apoplast route across the cortex

(ii) the symplast route across the cortex.

(a) true
(b) (i) water is considered to pass on the outside of the cell membrane
(ii) water passes through the cytoplasm of the cells through plasmodesmata.

Sample question and model answer

(a) The graph shows the oxygen dissociation curve for human haemoglobin.

Note that haemoglobin is able to pick up a lot of oxygen, even at low partial pressure.

Use the information in the graph to help you answer the following questions.

(i) What is the advantage of haemoglobin as a respiratory pigment when oxygen in the air is at the low partial pressure 6 kPa? [1]

Even at a low partial pressure a lot of oxygen (70%) is taken up by the haemoglobin of a red blood cell.

The fact that haemoglobin is able to carry oxygen is important. However, it is just as important that the oxygen is off-loaded at tissues needing it. This is only possible because carbon dioxide is found at the tissues.

(ii) Explain the effect on the oxygen dissociation curve of a high partial pressure of carbon dioxide at a muscle. [2]

The curve is moved to the right and down so that oxygen is released.

(iii) Fetal haemoglobin has a greater affinity for oxygen than maternal haemoglobin. Draw a curve on the graph to show the oxygen dissociation curve for fetal haemoglobin. [1]

See graph opposite.

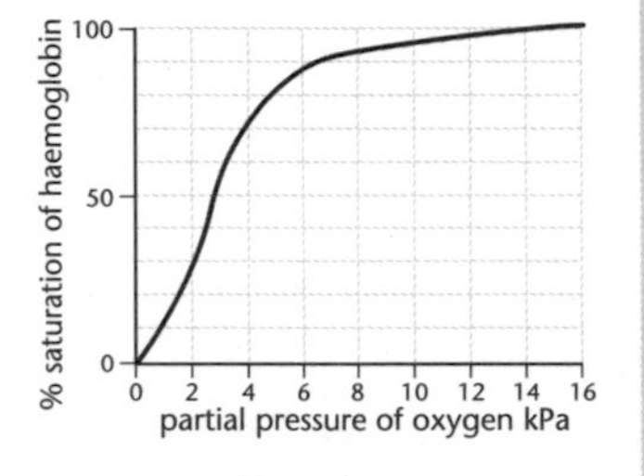

(b) The diagram shows **one** stage in the cardiac cycle.

Always look for the valves. If the heart valve is open then the chamber behind it is contracting.

(i) Which stage of the cardiac cycle is shown in the diagram? Give **two** reasons for your answer. [3]

atrial systole
the atrioventricular valves are open/blood flows through the atrioventricular valves,
semi-lunar valves are closed.

(ii) Write an X in one chamber to show the position of the atrioventricular node (AVN). [1]

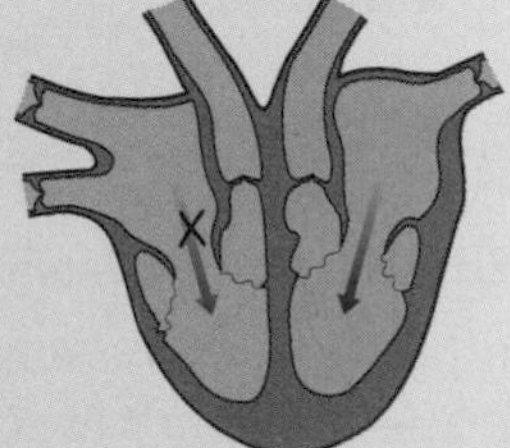

(iii) How does the AVN stimulate the contraction of the ventricles? [1]

Passes electrical impulses to Purkinje tissue/Bundle of His.

Practice examination questions

1 The diagram shows a capillary bed in the upper part of the skin. The arteriole is constricted.

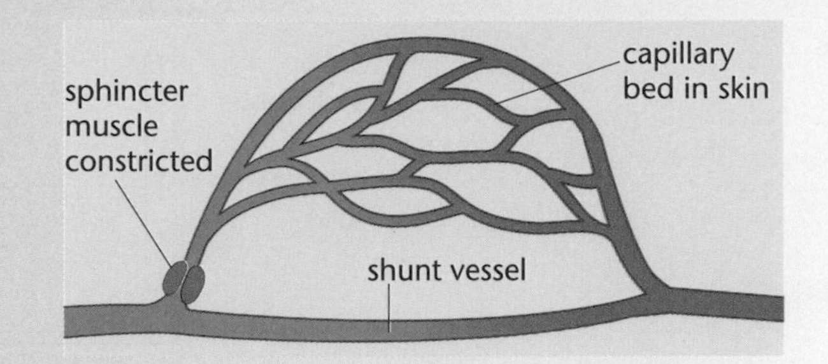

Use the information in the diagram and your own knowledge to answer the questions below.

(a) As a result of arteriole constriction, to where would the blood flow? [1]

(b) Explain how this would help maintain the body temperature. [4]

2 The table shows data about a person's heart before and after a training programme.

	Before training	***After training***
heart stroke rate	90 ml	120 ml
heart rate at rest	75 bpm	60 bpm
maximum heart rate	170 bpm	190 bpm

(a) Over a five minute period at rest before training, the cardiac output of the person was 33.75 litres.

How much blood would leave the heart, during the same time, whilst the person was at rest, after training? [2]

(b) After training, the maximum heart rate increased by 20 bpm. Explain the advantage of this increase to an athlete. [4]

(c) After training there are other changes in the body.

Explain:

(i) **two** changes which would improve the efficiency of the respiratory system. [2]

(ii) the effect of training on the muscles. [2]

Practice examination questions (continued)

3 The diagram shows a freshly cut, leafy shoot attached to a potometer. This was used to measure the amount of water taken up by the shoot under different conditions.

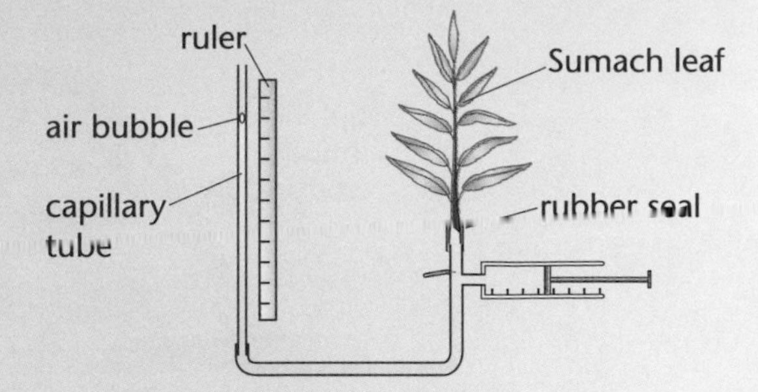

(a) What assumption must be made when using this apparatus to measure the rate of transpiration? [1]

(b) An air-lock can occur in the shoot which prevents water uptake.

(i) In which plant tissue could an air-lock occur? [1]

(ii) Describe the practical details by which a student could make sure that there was no air-lock in the shoot. [2]

(c) The radius of the capillary tube of the potometer was 1 mm. When a Sumach leaf was measured the air bubble moved 32 mm in one minute. Calculate the volume of water in mm^3 which would be taken up by the leaf in one hour under the same environmental conditions. [3]

4 *Agave americana* is a xerophytic plant which grows in the deserts of Mexico.

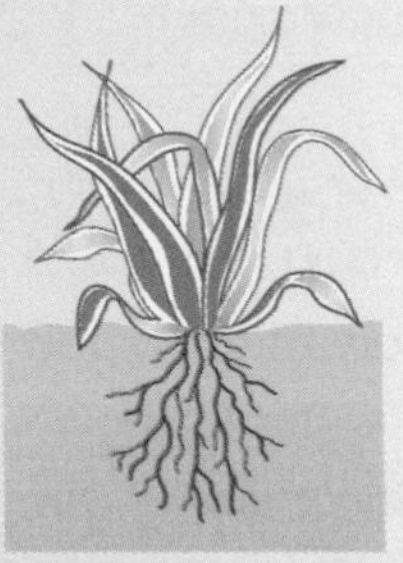

Agave americana

Suggest **three** ways in which the plant is adapted to survive periods of very low rainfall. [3]

5 The diagram shows nerves linking the medulla oblongata with the heart.

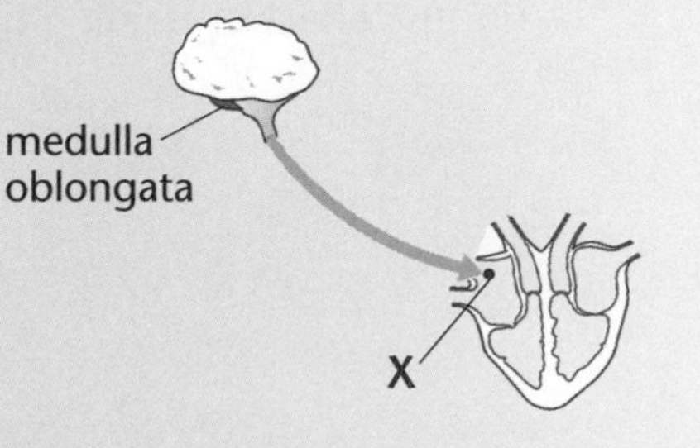

(a) Name part X. [1]

(b) What effect do the following have on the heart:

(i) vagus nerve

(ii) sympathetic nerve

(iii) adrenaline? [3]

Practice examination questions (continued)

6 The diagram shows an aphid feeding on a plant. The sharp stylet is inserted into the phloem tissue which supplies the aphid with sucrose, plus organic and inorganic ions.

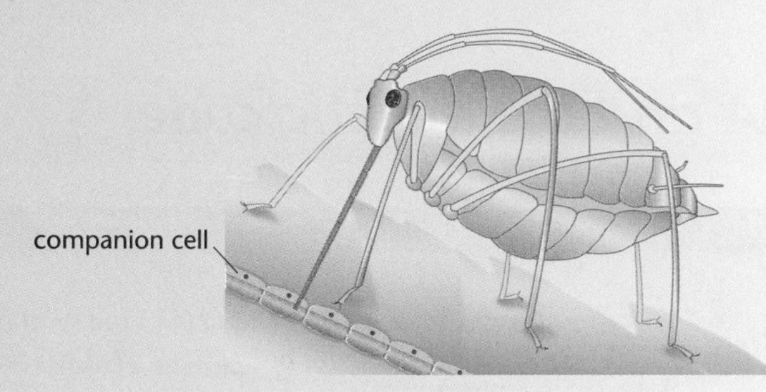

(a)

(i) Name the phloem cell X from which the aphid obtains sucrose. [1]

(ii) Cell X does not have a nucleus or ribosomes, but still contains enzymes. Explain how this is possible. [3]

(b)

(i) Feeding aphids obtain the contents of the phloem without any sucking action being necessary. What does this indicate about the transport of substances through the phloem? [3]

(ii) Scientists investigated phloem contents by anaesthetising feeding aphids, then cutting their bodies from their stylets. Phloem contents oozed from the cut end of each stylet. The phloem contents were tested using iodine and heating with Benedict's solution.

	Tested with iodine	*Heated with Benedict's solution*
Contents of phloem	brown colour	brick-red colour

Referring to the results of the tests, explain what the scientists found out about the phloem contents using this method. [4]

(iii) Hot-wax ringing is a technique where hot wax is poured around a stem. This technique was used with the aphid method described in (ii). Radioactive carbon dioxide was supplied to one leaf so that a radioactive carbohydrate was made.

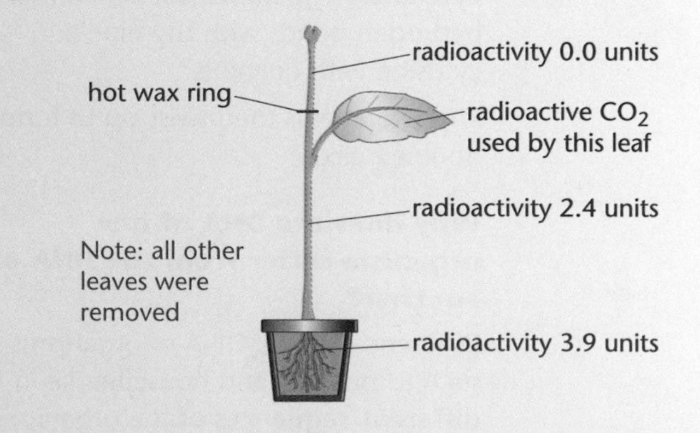

Explain the effect of hot-wax ringing on the phloem tissue. [3]

Chapter 6

Genes and cell division

The following topics are covered in this chapter:

- *DNA and the genetic code*
- *Cell division*
- *Gene technology*

6.1 DNA and the genetic code

After studying this section you should be able to:

- *recall the structure of DNA*
- *outline the roles of DNA and RNA in the synthesis of protein*
- *use organic base codes of DNA and RNA to identify amino acid sequences*

LEARNING SUMMARY

Deoxyribonucleic acid (DNA) and chromosome structure

AQA	2.2 2.3
CCEA	1.1
EDEXCEL	1.2.10
OCR	2.1.2
WJEC	1.6

Each chromosome in a nucleus consists of a series of genes. A gene is a section of a chemical called **DNA** and each gene controls the production of a polypeptide important to the life of an organism. **Deoxyribonucleic acid (DNA)** is made up of a number of **nucleotides** joined together in a double helix shape.

You need to be aware that many nucleotides join together to form the polymer, DNA.

Each nucleotide consists of a phosphate group, a molecule of deoxyribose sugar and an organic base. Phosphate and pentose sugar units link to form the backbone of the DNA. Repeated linking of the monomer nucleotides forms the polynucleotide chains of DNA.

Each strand of DNA is said to be complementary to the other. **Examination tip:** be ready to identify one strand when given the matching complementary strand.

Each DNA molecule is made up of two polynucleotide chains. The two chains are held together by hydrogen bonds between the bases.

The organic base of each nucleotide can be any one of **adenine, thymine, cytosine** or **guanine**. Adenine forms hydrogen bonds with thymine and cytosine with guanine.

The two chains then twist up to form a double helix.

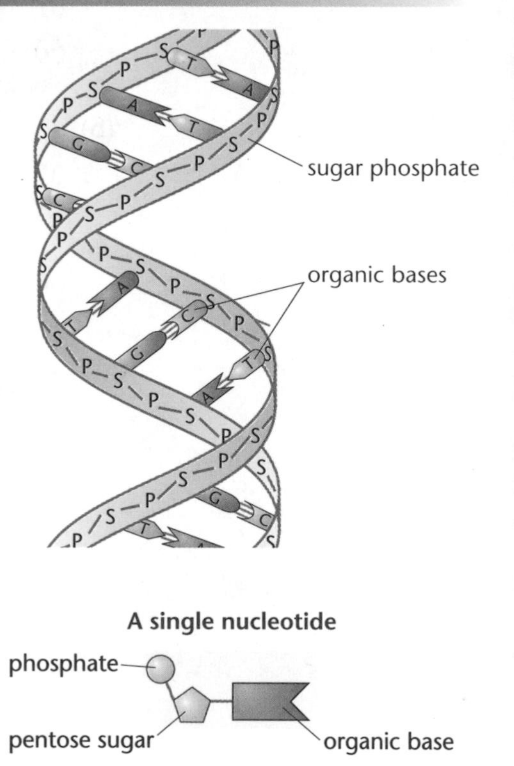

Why does the DNA of one organism differ from the DNA of another?

Differences in the DNA of organisms such as humans and houseflies lie in the **different sequences** of the organic bases. Each sequence of bases is a code to make a protein, usually vital to the life of an organism.

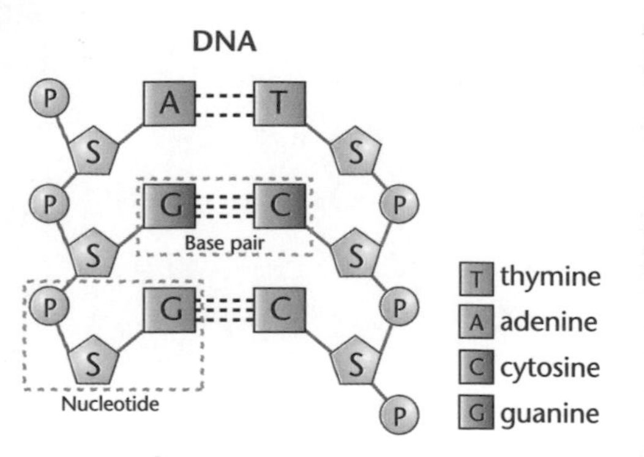

How does DNA control protein production?

AQA	2.2
CCEA	1.1
EDEXCEL	1.2.12 / 14
OCR	2.1.2
WJEC	1.6

Each different polypeptide is made of a specific order of amino acids and so DNA must code for this order.

KEY POINT

Each three adjacent bases code for one amino acid in the polypeptide. The genetic code is therefore a **triplet** code.

Use the key to identify the amino acids in the table opposite.

Amino acid	Abbreviation
alanine	Ala
arginine	Arg
asparagine	Asn
aspartic acid	Asp
cysteine	Cys
glutamine	Gln
glutamic acid	Glu
glycine	Gly
histidine	His
isoleucine	Iso
leucine	Leu
lysine	Lys
methionine	Met
phenylalanine	Phe
proline	Pro
serine	Ser
threonine	Thr
tryptophan	Trp
tyrosine	Tyr
valine	Val

The table below shows all the triplet sequences of organic bases found along DNA strands and the coding function of each.

Genetic code functions of DNA

	Second organic base								
	A		*G*		*T*		*C*		*Third organic base*
A	AAA	Phe	AGA		ATA	Tyr	ACA	Cys	**A**
	AAG		AGG	Ser	ATG		ACG		**G**
	AAT	Leu	AGT		ATT	stop	ACT	stop	**T**
	AAC		AGC		ATC	stop	ACC	Trp	**C**
G	GAA		GGA		GTA	His	GCA		**A**
	GAG	Leu	GGG	Pro	GTG		GCG	Arg	**G**
	GAT		GGT		GTT	Gln	GCT		**T**
	GAC		GGC		GTC		GCC		**C**
T	TAA		TGA		TTA	Asn	TCA	Ser	**A**
	TAG	Ile	TGG	Thr	TTG		TCG		**G**
	TAT		TGT		TTT	Lys	TCT	Arg	**T**
	TAC	Met	TGC		TTC		TCC		**C**
C	CAA		CGA		CTA	Asp	CCA		**A**
	CAG	Val	CGG	Ala	CTG		CCG	Gly	**G**
	CAT		CGT		CTT	Glu	C CT		**T**
	CAC		CGC		CTC		CCC		**C**

Do not learn all of the triplet codes. Be ready to use the supplied data in the examination. You will be given a key of different codes and functions.

If you are given a table of codes check them carefully. If the bases are from mRNA then there will be uracil in the table.

Each triplet code is **non-overlapping**. This means that each triplet of three bases is a code, then the next three, and so on along the DNA.

- AAA codes for the amino acid phenylalanine
- GAG codes for the amino acid leucine
- GAC codes for the amino acid leucine

There are more triplet codes than there are amino acids. This is known as the **degenerate code**, because an amino acid such as leucine can be coded for by up to six different codes. Some triplets do not code for amino acids but mark the beginning or end of polypeptides. They are **stop or start** triplets.

RNA and protein synthesis

DNA is found in the nucleus but proteins are made on ribosomes in the cytoplasm. Therefore a messenger is needed to transfer the code. This messenger is a molecule called ribonucleic acid (**RNA**).

RNA is a nucleic acid, made up of nucleotides like DNA but it has some important differences:

DNA	*RNA*
Two polynucleotide strands	One polynucleotide strand
Contains adenine, cytosine, guanine and thymine	Contains adenine, cytosine, guanine and the base uracil instead of thymine
Contains deoxyribose sugar	Contains ribose sugar

This type of RNA is called messenger RNA (mRNA) because it carries the message out of the nucleus. There are two other types of RNA called rRNA and tRNA.

Messenger RNA is formed in the nucleus by making a complementary copy of the DNA coding for the polypeptide.

Here is an example of a coding strand of DNA:

DNA	A A A G A G G A C A C T	*(coding strand)*
mRNA	U U U C U C C U G U G A	*(messenger RNA)*

guanine (G) on DNA codes for cytosine (C) on mRNA

cytosine (C) on DNA codes for guanine (G) on mRNA

thymine (T) on DNA codes for adenine (A) on mRNA

adenine (A) on DNA codes for uracil (U) on mRNA

Progress check

(a) Name the parts of a nucleotide.

(b) (i) By which bonds do the two strands of DNA link together?

(ii) How would these bonds be broken in the laboratory to produce single strands of the DNA?

(c) Which organic base is found in DNA but not in RNA?

(a) pentose sugar, phosphate and organic base. The organic base may be thymine, adenine, cytosine or guanine

(b) (i) hydrogen bonds (ii) heat

(c) thymine

6.2 Cell division

LEARNING SUMMARY

After studying this section you should be able to:

- *describe and explain the semi-conservative replication of DNA*
- *understand that DNA must replicate before cell division can begin*
- *recall the purpose of mitosis and meiosis*
- *recognise each stage of the cell cycle and cell division by mitosis*

How do cells prepare for division?

AQA	2.5
CCEA	1.1
EDEXCEL	1.2.11
OCR	2.1.2

Before cells divide they must first make an exact copy of their DNA by using a supply of organic bases, pentose sugar molecules and phosphates. The method by which DNA is copied is called **semi-conservative replication**. The diagram (right) shows this taking place.

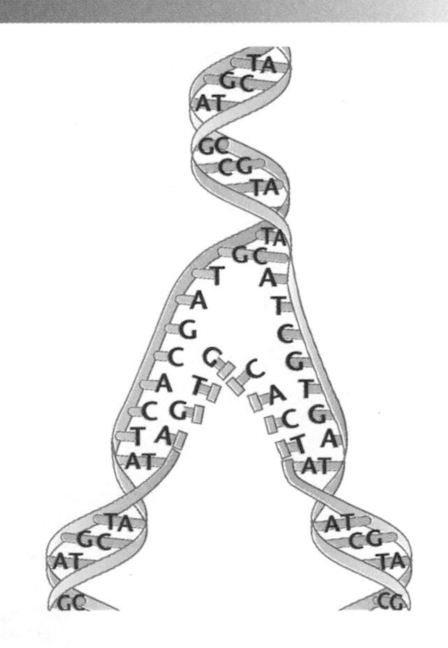

Remember that as the DNA unwinds each single strand is a **complement** to the other. This means that each has the **matching** series of organic bases.

- The DNA begins to unwind under the influence of the enzyme DNA helicase.
- Hydrogen bonds between the two chains then break and the two strands separate.
- Each complementary strand then acts as a template to build its opposite strand from free nucleotides.
- The enzyme DNA polymerase joins the nucleotides together. This process results in the production of two identical copies of double-stranded DNA.

Evidence for the semi-conservative replication of DNA

> Exam questions are often based on experimental data. Apply your knowledge of principles learned during the course and this will be helpful.

The first real evidence came from the results of an experiment carried out by two researchers, Meselson and Stahl.

Bacteria were cultured with a heavy isotope of nitrogen located in the organic bases of their DNA.

The bacteria were then supplied with bases containing the normal light nitrogen atoms. They replicated their DNA using these bases. Their population increased.

Each molecule of DNA of the next generation had one strand containing heavy nitrogen and one strand containing light nitrogen. The mass of the DNA was therefore midway between the original heavy form and normal light DNA.

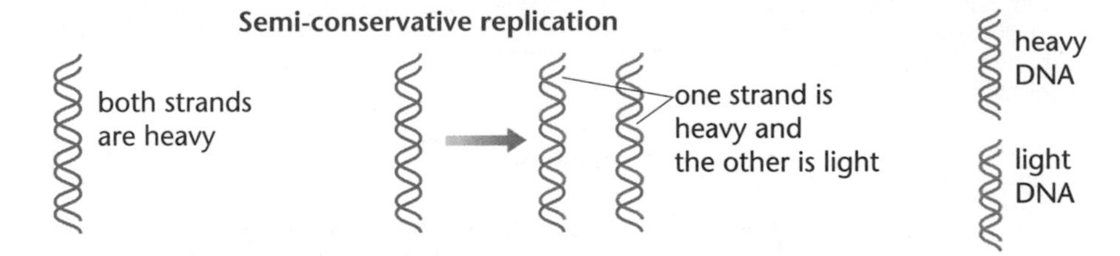

> Make sure that you can predict what would happen if the bacteria replicated again.

Semi-conservative therefore means that as DNA splits into its two single strands, each of the new strands is made of newly acquired bases. The other strand, part of the original DNA, remains.

Types of cell division

AQA	2.5
CCEA	1.1
EDEXCEL	1.2.11
OCR	2.1.2

Cells divide for the purposes of growth, repair and reproduction. Not all cells can divide but there are two ways in which division may occur: **mitosis** and **meiosis**.

In most organisms, the chromosomes in each cell can be arranged in pairs. Each cell therefore has two copies of each gene.

When the chromosomes are in pairs, the cell is said to be **diploid** and the pairs are called homologous pairs.

> Be careful of the spelling of mitosis and meiosis; examiners may penalise you if they are spelt incorrectly.

Mitosis produces genetically identical copies of cells with the same number of chromosomes for growth and repair.

Meiosis produces cells that have half the number of chromosomes, one from each pair. These cells are **haploid** and are used as gametes.

Meiosis introduces variation because the pairs of chromosomes can be split up in many different ways. This is called independent assortment.

The cell cycle and mitosis

AQA	2.5
CCEA	1.1
EDEXCEL	2.3.6
OCR	1.1.3
WJEC	1.7

The length of time between a cell being formed and it dividing is called the **cell cycle**. This can be divided up into a number of different phases:

- G1 phase – the cell grows making new proteins and more organelles
- S phase – the DNA of the chromosomes is replicated by semi-conservative replication
- G2 – more organelles are made and a spindle forms
- M phase – this is mitosis involving the separating of the genetic material into two nuclei
- C phase – cytokinesis, where the cell divides into two.

The cell cycle
G_2
prophase
metaphase
anaphase
telophase
C
G_1
interphase
S

Remember that DNA replication takes place before cell division in **interphase** (see page 85). This is not an integral phase of mitosis or meiosis.

The length of the cell cycle varies between different types of cells. Some cells never divide once formed but cells of the bone marrow divide about every eight hours. Interphase usually takes up 95% or more of the whole cell cycle.

Be ready to analyse photomicrographs of all phases of mitosis. If you can spot 10 pairs of chromosomes at the end of telophase, then this is the original diploid number of the parent cell.

Interphase is the period of time between cell divisions. It is made up of G1, S and G2 phases. Once cells start mitosis, they all go through a similar sequence of events. This is shown in the diagrams.

1 Prophase

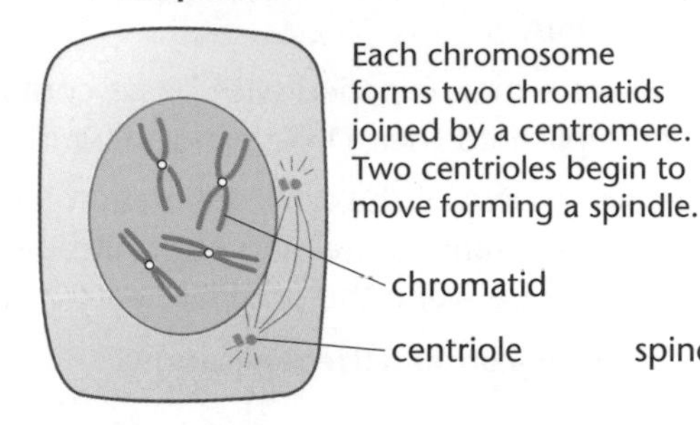

Each chromosome forms two chromatids joined by a centromere. Two centrioles begin to move forming a spindle.

2 Metaphase

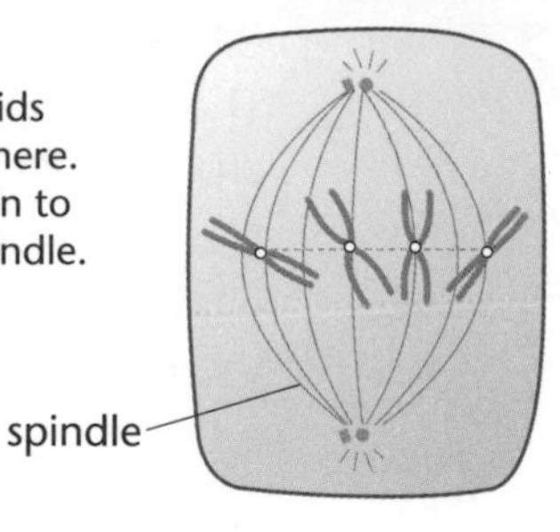

The chromatids, still joined by a centromere move to the middle of the cell. Each of the two chromatids has identical DNA to the other.

3 Anaphase

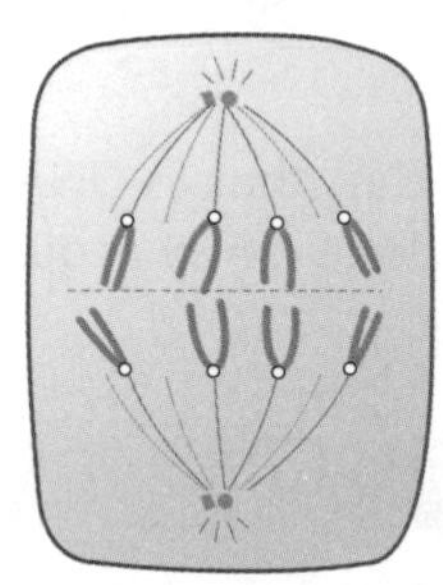

The spindle fibres join to the centromeres. The spindle fibres shorten and the centromeres split. The separated chromatids are now chromosomes.

4 Telophase

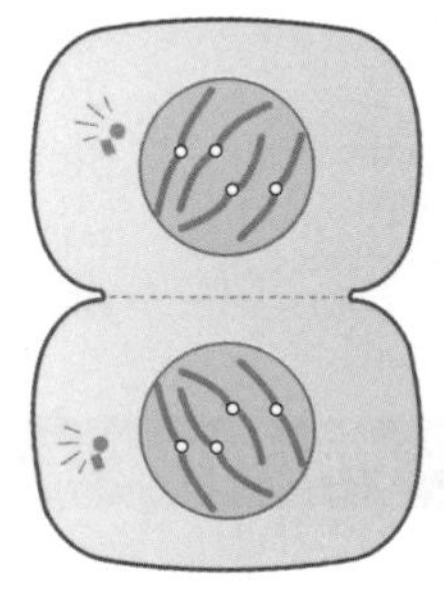

Identical chromosomes move to each pole. The nuclear membrane re-forms. The cell membrane narrows at the middle and two daughter cells are formed.

6.3 Gene technology

After studying this section you should be able to:

- *define different types of stem cells*
- *understand the process of electrophoresis and recall its applications*

LEARNING SUMMARY

Stem cells

EDEXCEL 2.3.11
OCR 1.1.3

When cells have differentiated and become specialised they lose their ability to divide to form other types of cells. They also can only divide a limited number of times. **Stem cells** are undifferentiated cells that can divide to form different types of cells. There are different types of stem cells:

- Embryonic stem cells are **totipotent**. This means that they can form any type of cell.
- Later in the embryo and in the adult, the stem cells are **pluripotent**. This means that they can form certain types of cells.

Stem cells have the potential to be used in many types of medical therapies but the use of embryonic stem cells, in particular, has raised a number of ethical issues.

Electrophoresis and genetic fingerprinting

CCEA 1.3

Restriction endonucleases enzymes can be used to cut up an organism's DNA.

- DNA sections are put into a well in a slab of agar gel.
- The gel and DNA are covered with buffer solution which conducts electricity.
- Electrodes apply an electrical field.

- Phosphate groups on DNA are negatively charged causing DNA to move towards the anode.
- Smaller pieces of DNA move more quickly down the agar track, larger ones move more slowly, leading to the formation of bands.

a view looking down on the agar slab

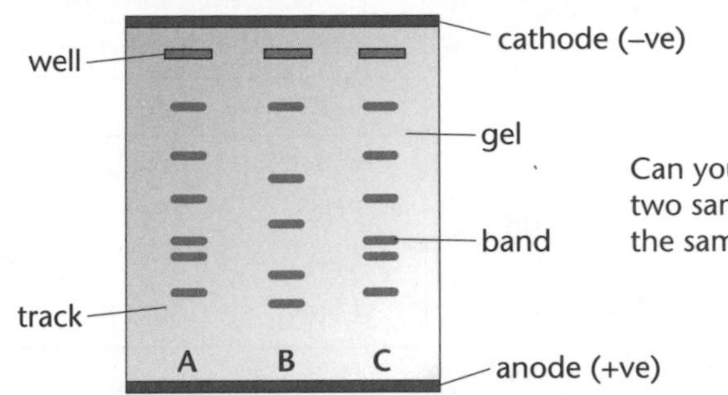

Can you spot which two samples were from the same person?

Genetic fingerprinting is applied to organisms other than humans. Illegal egg collectors have been successfully prosecuted when egg DNA has been compared against the DNA profile of parent birds.

Electrophoresis has many applications. DNA is highly specific so the bands produced using this process can help with identification. In some crimes DNA is left at the scene. Blood and semen both contain DNA specific to an individual. DNA evidence can be checked against samples from suspects. This is known as **genetic fingerprinting**. Genetic fingerprinting can be used in paternity disputes. Each band of the DNA of the child must correspond with a band *either* from the father or from the mother.

Progress check

A length of DNA was prepared and then electrophoresis was used to separate the sections. The statements below describe the process of electrophoresis but they are in the wrong order. Write the letters in the correct sequence.

A electrodes apply an electrical field
B DNA sections are put into a well in a slab of agar gel
C smaller pieces of DNA move more quickly down the agar track with larger ones further behind
D the gel and DNA are then covered with buffer solution which conducts electricity
E restriction endonucleases can be used to cut up the DNA

E B D A C

Sample questions and model answers

Always remember to read all information given. You must be aware of what the examiner is testing. Try to link the question with the part of the specification the topic comes from.

You will not be expected to remember all the genetic code! In this question you are given data. Make sure you have revised all key words so that you can apply the principles you have learned to the data.

Exam boards never give the full name of amino acids. There will always be a key.

The table below shows some mRNA codons and the amino acids which are coded by them.

first position	*second position*				third position
	U	*C*	*A*	*G*	
U	Phe	Ser	Tyr	Cys	U
	Phe	Ser	Tyr	Cys	C
	Leu	Ser	stop	stop	A
	Leu	Ser	stop	Trp	G

Key to amino acids

Ser – serine
Phe – phenylalanine
Leu – leucine
Tyr – tyrosine
Trp – tryptophan
Cys – cysteine

Use the information in the table to help you answer the following questions.

There are two codes to choose from.

1

(a) Give a sequence of mRNA bases which would code for leucine. [1]

UUA or UUG

(b) What does the mRNA base sequence UAC code for? [1]

Tyrosine

Follow the table headings to indicate U (first base), A (second base) and C (third base).

This can be tricky! Remember that you need to work backwards. Given mRNA you know that it is coded for by one strand of DNA. Work out one DNA strand then use T – A, and G – C links.

2

The mRNA sequence UCA codes for serine. Work out the base pairs on the DNA. [3]

UCA is coded for by these bases: AGT
AGT links to the bases TCA
So the DNA is AGT
TCA

If you failed to learn the definition of the degenerate code then you would fail to apply it to this data.

3

Use evidence from the table to show that serine is an example of the degenerate code. [1]

It is coded for by four different base sequences.

4

UAG codes for 'stop'. Explain the effect of the 'stop' code during the process of protein synthesis. [2]

It is responsible for the polypeptide being terminated which allows it to leave the ribosome once all the amino acids have been linked.

Practice examination questions

1 The diagram below shows a stage in the process of mitosis.

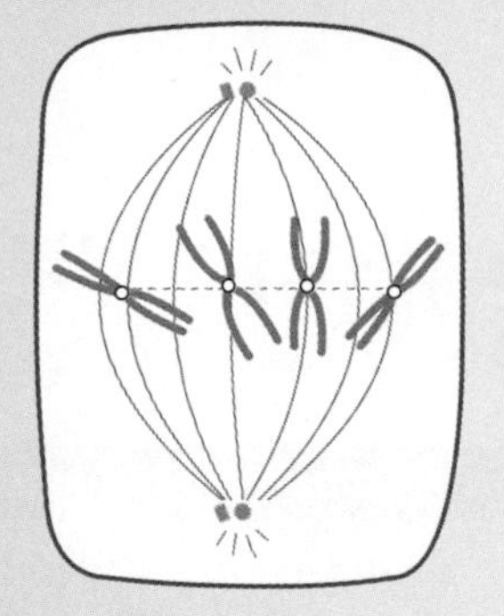

(a) Give the stage of mitosis shown. [1]

(b) How many chromosomes would there be in the daughter cells? [2]

2 The table below shows the relative organic base proportions found in human, sheep, salmon and wheat DNA.

	Proportion of organic bases in DNA (%)			
Organism	***Adenine***	***Guanine***	***Thymine***	***Cytosine***
human	30.9	19.9	29.4	19.8
sheep	29.3	21.4	28.3	21.0
salmon	29.7	20.8	29.1	20.4
wheat	27.3	22.7	27.1	22.8

(a) Refer to the proportion of organic bases in salmon DNA to explain the association between specific bases. [2]

(b) Suggest a reason for the small difference in proportion of the organic bases adenine and thymine in sheep. [1]

(c) All species possess adenine, guanine, thymine and cytosine in their DNA. Account for the fact that each species is different. [2]

3 It was suspected that a person had taken an egg from the nest of a rare bird. DNA samples were taken from the egg and both parent birds. The DNA profiles shown below were made using electrophoresis.

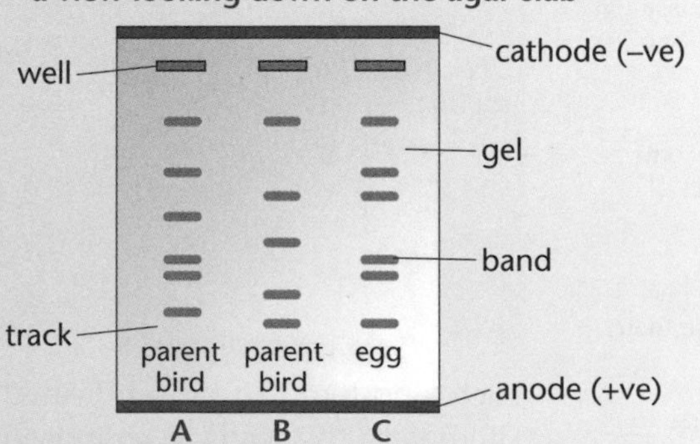

(a) Which type of enzyme is used to cut DNA before electrophoresis? [1]

(b) Do the electrophoresis results suggest that the egg was taken from the nest? Give a reason for your answer. [1]

(c) Suggest TWO other applications for electrophoresis. [2]

Chapter 7

Classification and biodiversity

The following topics are covered in this chapter:

- *Classification*
- *Biodiversity*
- *Maintaining biodiversity*

7.1 Classification

After studying this section you should be able to:

LEARNING SUMMARY

- *understand why and how organisms are classified*
- *describe the binomial system of naming organisms*
- *describe some of the more recent techniques used to classify organisms*

Classifying and naming

AQA 2.8
CCEA 2.3
OCR 2.3.2
WJEC 2.1

It is estimated that there are more than 10 million different kinds of living organisms alive on Earth. The study of the range of different organisms is called **systematics**. For centuries, scientists have tried to classify organisms into groups and give them names.

KEY POINT

The system of classifying organisms into groups is called **taxonomy** and the system of naming is called **nomenclature**.

Organisms are put into groups based on the largest number of common characteristics. Firstly, they are put into one of five large groups called kingdoms.

The table shows the characteristics of organisms in the five kingdoms:

Try this!

Pretty **P**olly **F**inds **P**arrots **A**ttractive.

It will help you to remember the kingdoms!

Most examination boards test knowledge of characteristics across the kingdoms.

Information will usually be given for any subgroup tasks.

The table includes some of the main features of each kingdom.

Prokaryotae and Protoctista tend to give more of a challenge than some of the other kingdoms, and are examined more often.

→ *increase in complexity*				
Prokaryotae	***Protoctista***	***Fungi***	***Plantae***	***Animalia***
very simple cells with few organelles no membrane-bound organelles if there are flagellae, then not 9 + 2 system of microtubules DNA in strands, no true nucleus e.g. bacteria and cyanobacteria	unicellular cells with membrane-bound organelles some are photosynthetic, but many have heterotrophic nutrition reproduction usually involves fission e.g. algae and protozoa	heterotrophic nutrition some saprotrophic, some parasitic consists of thread-like hyphae, chitin cell walls many nuclei in hyphae, not in one per cell organisation reproduction involves the production of spores	multicellular organisms which are photosynthetic cells have cellulose cell wall, sap vacuole and chloroplasts reproduce by seeds, or by spores, some sexual, some asexual.	multicellular organisms which are heterotrophic no cell walls no sap vacuoles

Each kingdom can be subdivided into a number of progressively smaller groups. Ultimately, this leads to an individual type of organism, a **species**.

The seven groups can be difficult to remember. Try the easy way!

King Penguins Climb Over Frozen Grassy Slopes

The first letter of each word will help you remember. It is a mnemonic which is an excellent strategy to aid recall.

The hierarchy of the groups is shown below.

	Example 1	*Example 2*
Kingdom	Animalia	Animalia
Phylum	Chordata	Arthropoda
Class	Mammalia	Insecta
Order	Primates	Lepidoptera
Family	Hominidae	Pieridae
Genus	Homo	Pieris
Species	sapiens	brassica

What is a species?

Remember, similar organisms such as horses and donkeys can mate to produce mules, but mules are infertile.

The smallest classification group is the species.

KEY POINT

Organisms are members of the same species if they can breed together to produce **fertile offspring**.

Naming organisms

Most organisms are known by common names. We use these names all the time. The problem is that an organism may be known by different names or sometimes different organisms can have the same common name. A scientist called Linnaeus introduced a scientific naming system so that each species could have a unique scientific name. This system is called the **binomial system**.

The name consists of two parts, the name of the genus written with a capital letter and the name of the species, written in lower case.

Notice that binomial names are always typed in italics or underlined if handwritten.

Panthera leo

Panthera tigris

Modern classification techniques

AQA	2.8
CCEA	2.3
EDEXCEL	2.4.16
OCR	2.3.2
WJEC	2.1

For centuries, organisms were classified according to observable physical features. This might be the structure of an animal's skull or the shape of a plant's leaves. Scientists can now use a range of different techniques.

- Microscopic structure – modern electron microscopes have shown that bacteria and other single-celled organisms such as amoeba have completely different cell structure and so they are now put into different kingdoms: *Prokaryotae* and *Protoctista*.

- Genetic differences – it is now possible to compare the DNA base sequences of different organisms. This can be done by a process called **DNA hybridisation**. Short, single-stranded sections of DNA are produced from one species and the extent to which they bind with DNA from another species is measured.
- Biochemical differences – the occurrence of different biochemical molecules is often a good indicator of relationships. It is also possible to work out and compare the amino acid sequence of common proteins.
- Immunological evidence – antibodies against human proteins can be made and then tested against proteins from other animals. The more effective the antibodies are, then the closer the evolutionary relationship is between humans and the other animal.

Immunological results give the following % similarities to man:
chimpanzee 97%,
gibbon 92%,
lemur 37%,
pig 8%

The aim of modern classification systems is to use a range of techniques to produce a system that classifies organisms based on their **evolutionary relationships**.

KEY POINT: A system based on evolutionary relationships is called a **phylogenetic system.**

7.2 Biodiversity

LEARNING SUMMARY

After studying this section you should be able to:

- *understand what is meant by the term biodiversity*
- *explain why there is so much biodiversity on Earth*

What is biodiversity?

AQA	2.11
CCEA	2.3
Edexcel	2.4.13
OCR	2.3.2

Biodiversity is a measure of the variation between different living organisms. It could be:

- **genetic diversity** – the differences between the genes in a species
- **species diversity** – the number of different species in a community
- **ecosystem or habitat diversity** – the variety of different areas where organisms can live.

A few definitions

A pond is a habitat containing populations such as stickleback and pondweed. All the organisms make up the community and the organisms, the water, the mud, etc. are an ecosystem.

A **habitat**: an area in which an organism lives.
A **population**: all the organisms of one species living in a habitat.
A **community**: all the different species (different populations) living in a habitat.
An **ecosystem**: all the living organisms (biotic) and inorganic parts (abiotic) of a habitat.
A **niche** is where an organism lives and its role in that ecosystem.

Estimating populations

Point quadrat

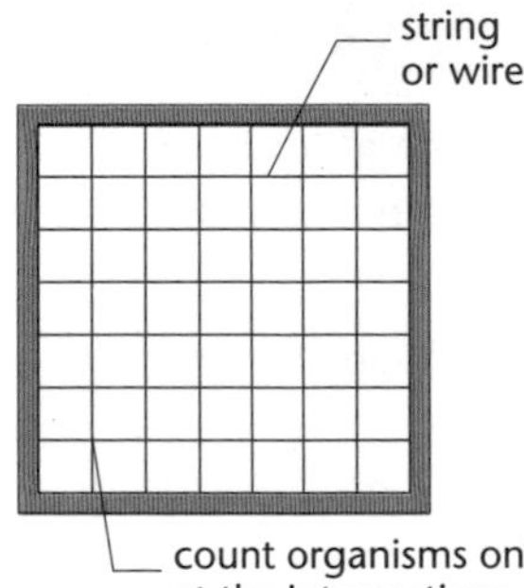

Often a full count of organisms in an ecosystem is not possible because of the size of the ecosystem. A **sampling technique** is used which requires a **quadrat**. This is a small area enclosed by wire or wood, around 0.25 m^2. When placed in the ecosystem the organisms inside the area can be counted, as well as the **abiotic factors** which influence their distribution. Ecologists use units to measure organisms within the quadrats. Frequency (f) is an indication of the presence of an organism in a quadrat area. This gives no measure of numbers. However the usual unit is that of density, the numbers of the organism per unit area. Sometimes percentage cover is used, an indication of how much of the quadrat area is occupied.

Consider a survey of two species *Taraxacum officinale* (dandelion) and *Plantago major* (Great plantain) of the lawn habitat shown below.

A simplified results table

Quadrat no.	Dandelion
1	2
2	12
3	15
4	3
5	4
6	8
7	7
8	10
9	9
10	15

mean = 8.5 per quadrat
Dm^{-2} = 34

Lawn habitat

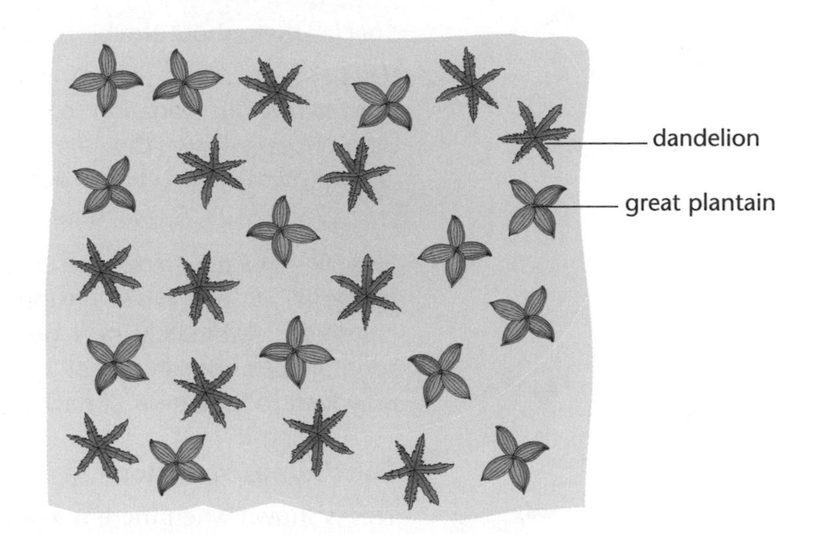

It is important to use a suitable technique when surveying with quadrats. When you observe a habitat which appears **homogeneous** or **uniform**, like a field yellow with buttercups, then you should use **random quadrat placement**. The area should be gridded, numbers given to each sector of the grid, then a random number generator used. The probability of the numbers in one quadrat representing a field would be very low. In practice the **mean** numbers from large numbers of quadrats do represent the true numbers in a habitat.

The index of biodiversity

This is used as a measure of the range and numbers of species in an area.

$$\text{index } d = \frac{N(N-1)}{\Sigma\, n(n-1)}$$

N = total no. of all individuals of all species in the area
n = total no. of individuals of one species in an area
Σ = the sum of

Consider this example of animals in a small pond

crested newt	8
stickleback	20
leech	15
great pond snail	20
dragonfly larva	2
stonefly larva	10
water boatman	6
caddisfly larva	30
	N = 111

$$d = \frac{111 \times 110}{(8 \times 7) + (20 \times 19) + (15 \times 14) + (20 \times 19) + (2 \times 1) + (10 \times 9) + (6 \times 5) + (30 \times 29)}$$

$$d = \frac{12\,210}{2018} \quad \text{so } d = 6.05$$

In another pond there were:

crested newt	45
stickleback	4
leech	18
great pond snail	10

d = 2.6

Look at both indices. 6.05 is an indicator of greater diversity. The higher number indicates greater diversity.

Sources of biodiversity

AQA	2.1
CCEA	2.3
EDEXCEL	2.4.14–15
OCR	2.3.3

What is variation?

Species throughout the biosphere differ from each other.

Variation describes the differences which exist in organisms throughout the biosphere. This variation consists of differences **between** species as well as differences **within** the same species. Each individual is influenced by the environment, so this is another source of variation.

genotype + environment = phenotype

The alleles which are expressed in the phenotype can only perform their function efficiently if they have a supply of suitable substances and have appropriate conditions. Ultimately new genes and alleles have appeared by mutation. The spontaneous appearance of **advantageous** new mutations is also possible. This may lead to formation of new species.

Remember that a species shows continuous variation when there are small incremental differences, e.g. height of people in a town. Beginning with the smallest and ending with the tallest there would probably be at least one person at each height, at 1cm increments. A smooth gradation of differences!

Height of person

Continuous variation

This is shown when there is a range of **small incremental differences** in a feature of organisms in a population. An example of this is height in humans. If the height of each pupil in a school is measured then from the shortest pupil to the tallest, there are very small differences across the distribution. This is shown by the graph below which shows smooth changes in height across a population. This type of variation is shown when features are controlled by **polygenic inheritance**. A number of genes **interact** to produce the expressed feature.

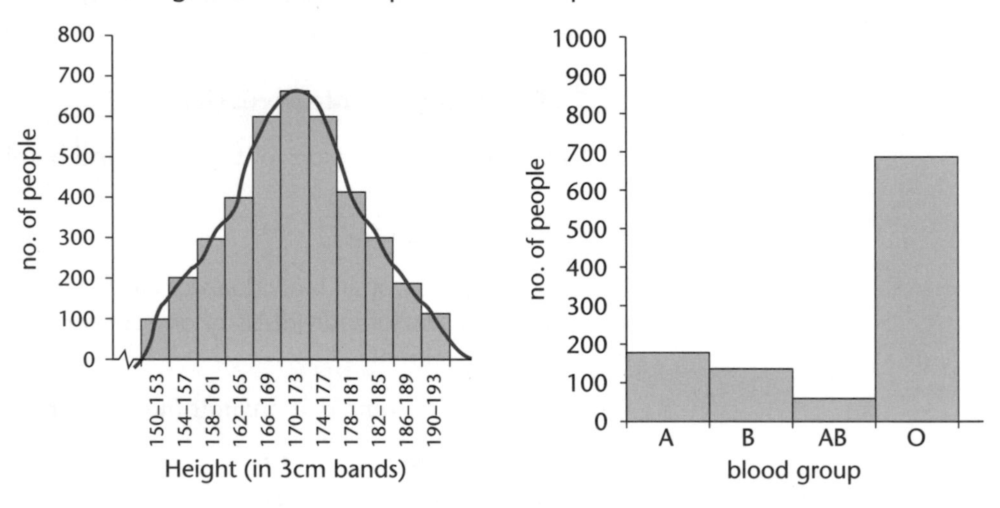

Discontinuous variation

This is shown when a characteristic is expressed in discrete categories. Humans have four discrete blood groups, A, B, AB or O. There are no intermediates, the differences are clear cut!

Mean and standard deviation

The variation shown by most populations can be described graphically. The variable usually forms a bell-shaped curve called a **normal distribution**.

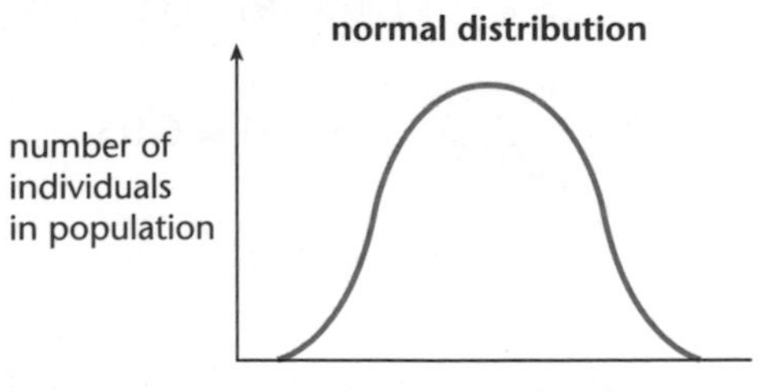

In a normal distribution, 68% of the values are within one standard deviation either side of the mean and 95% are within two.

The **mean** is the value at the peak of the curve. The **standard deviation** describes how the values are spread out. The higher the standard deviation, then the more variation there is.

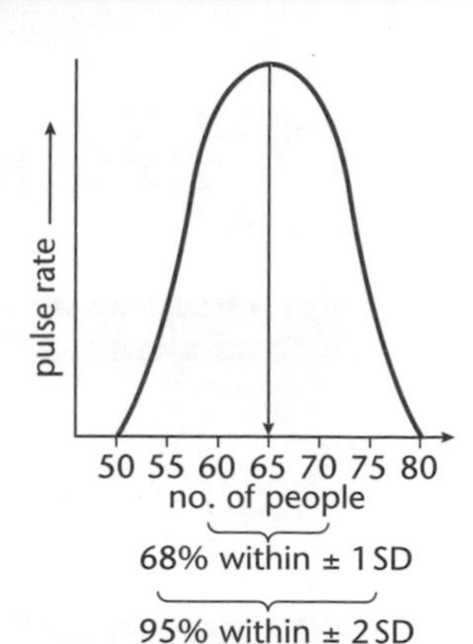

Standard deviation

Why do organisms show variation?

The many different habitats on Earth provide different conditions for organisms to live in. There may be variations in a number of different factors that can affect organisms:

- **climatic** factors, e.g. temperature, water availability, light
- **edaphic** (soil) factors, e.g. pH, minerals and oxygen content
- **biotic** factors, e.g. competition.

Organisms have become adapted to living in a particular habitat with a particular set of factors. These factors may be behavioural, physiological and anatomical. The table below shows some examples.

Type of organism	*Adaptation*	*Behavioural, physiological or anatomical*
cacti	leaves are spines to reduce water loss	anatomical
birds	parental care shown to young	behavioural
carnivores and herbivores	particular dentitions adapted to deal with specialised diets	anatomical
llamas	production of haemoglobin molecules with a higher affinity for oxygen	physiological

How is variation produced?

The vast biodiversity of organisms on Earth has been produced by the process of **natural selection** over millions of years. These are the key features of natural selection:

The survival of the best adapted organisms is often called 'survival of the fittest'.

Populations of organisms show variation and some of this variation can be inherited.

More organisms are born than can survive in a particular environment.

There must be a struggle for survival. Those organisms that are best adapted to the environment will survive.

The organisms that survive will breed and pass on their genes.

The theory of natural selection was first put forward by Charles Darwin in 1858 in his book *On the Origin of Species*. The small changes to populations that occur over long periods of time could result in the formation of new species. This is called **speciation**.

Recently, natural selection has been used to explain the development of:

- pesticide resistance in insects
- drug resistance in microorganisms.

7.3 Maintaining biodiversity

LEARNING SUMMARY

After studying this section you should be able to:

- *appreciate why it is important to maintain biodiversity*
- *describe some of the approaches used to try and maintain biodiversity*

Why is it important to maintain biodiversity?

CCEA 2.3
OCR 2.3.4

The human population of the Earth is rapidly increasing. This has resulted in increasing pollution and overexploitation of the Earth's natural resources. This has tended to reduce the biodiversity of many ecosystems and threaten entire habitats.

Species conservation involves managing the Earth's resources so that the biodiversity of animals and plants can be maintained. There are a number of reasons why people think that this is necessary.

- **Economic** – animal or plant species that are important to the economy of countries for food or raw materials may be lost.
- **Ecological** – the loss of species may have implications for the survival of other organisms.
- **Ethical** – many people think that we have no right to cause the extinction of other species.
- **Aesthetic** – many people like to enjoy the variety of organisms and habitats of the Earth.

Approaches to maintaining biodiversity

Conservation does not involve just preserving wildlife but requires active maintenance to counteract the damage being done by mankind. Many of the threats to habitats are changes that are occurring on a world-wide level. An example of this is global warming which may alter the climate of the whole planet. Successful conservation therefore requires action at **international**, **national** and **local** levels.

International conservation

It is often difficult but essential to get international agreement for conservation projects.

CITES

This is The Convention in International Trade in Endangered Species. It is an agreement between 172 different governments that restricts the trade of certain species. Lists of endangered species have been drawn up and roughly 5000 species of animals and 28 000 species of plants are protected by CITES against over-exploitation through international trade.

Rio Convention on Biodiversity

This is an international treaty that was adopted at the Earth Summit in Rio de Janeiro in June 1992. The Convention has three main objectives.

1. The conservation of biological diversity (or biodiversity).
2. The sustainable use of biological resources.
3. The fair and equitable sharing of benefits arising from genetic resources.

The ongoing arguments over whaling restrictions are a good example of how difficult it is to get binding international agreements.

National projects

Many countries have zoos or botanical gardens that have been popular with visitors for many years. Now many of these establishments are focusing more on the conservation of species rather than just entertaining the public. This has been encouraged by grants and charitable status.

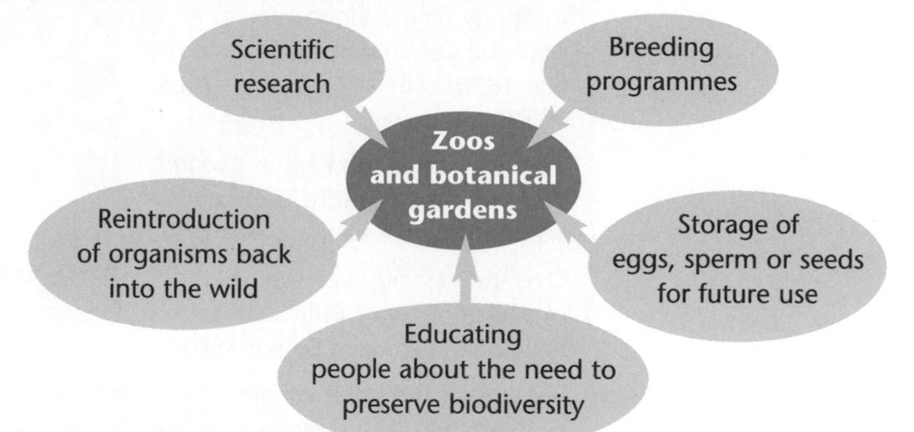

Although these zoos and botanical gardens are funded on a national basis, they still require international cooperation in terms of setting up breeding programmes and storing material. This is because it is vital to maintain genetic variety in order to prevent inbreeding.

Local conservation

Candidates for CCEA Biology should know details of some the strategies and initiatives specific to Northern Ireland.

It is now necessary for developers planning certain building projects to construct an Environmental Impact Statement before they ask for planning permission. The statement describes the likely significant effects of the development on the environment and biodiversity. The statement must be taken into account by the local planning authority before it may grant consent.

Sample question and model answer

Questions about natural selection always have similar mark schemes. Just make sure that you apply this mark scheme to the particular situation.

The quagga is an extinct mammal that looked like a zebra but had stripes only on its head, neck and forebody. The quagga lived in Africa until the last animal was shot in the late 1870s. The last specimen died in 1883 in a zoo.

The quagga, named *Equus quagga*, was originally thought to be a separate species to the plains Zebra.

Over the last fifty years or so, the markings of many individual zebras have been recorded by scientists.

Because of the great variation in stripe patterns, taxonomists were left with a problem. Was the quagga a separate species or was it a type of zebra? The quagga was the first extinct creature to have its DNA studied. Recent research using DNA from preserved specimens has demonstrated that the quagga was not a separate species at all, but a variety of the plains zebra, *Equus burchelli.*

(a) What is the name of the genus that both the plains zebra and the quagga belong to? [1]

Equus

(b) What type of variation does the stripe pattern in zebras demonstrate? [1]

Continuous variation

(c) How could scientists use the DNA to investigate the relationship between the zebra and the quagga? [2]

Use DNA hybridisation.
The DNA is split into single strands and then scientists see how well it binds with zebra DNA strands.

(d) The plains zebra may have more stripes than the quagga because the stripes provide better camouflage on the plains.
How might this pattern have developed by natural selection? [3]

Zebras show variation and are born with different patterns of stripes.
More stripes make the zebra better camouflaged and so more likely to survive.
These zebras survive, breed and pass on the genes for more stripes.

Practice examination questions

1 The following key distinguishes between the five kingdoms.

1	Organisms without membrane bound organelles Organisms with membrane bound organelles	A GOTO 2
2	Organisms have hyphae Organisms do not have hyphae	B GOTO 3
3	Organisms unicellular or colonial Organisms not unicellular or colonial	C GOTO 4
4	Organisms multicellular and have thylakoid membranes in some cells Organisms multicellular and have no thylakoid membranes in any cells	D E

Name kingdoms A, B, C, D and E [5]

2 (a) The song-thrush (*Turdus ericetorum*) and mistle-thrush (*Turdus viscivorus*) are in the same family, *Turdidae.* Large sections of their DNA are common to both species. Complete the table to classify both organisms. [3]

	Mistle-thrush	***Song-thrush***
Kingdom		
Phylum	Chordata	Chordata
	Aves	
	Passeriformes	Passeriformes
Genus		
Species		
[1]	[1]	[1]

(b) How is it possible to find out if two female animals are from the same species? [2]

3 The following article appeared in a newspaper.

ATLANTIC COD IN DANGER OF EXTINCTION

The cod stocks of the Atlantic Ocean are in urgent need of protection. Deep-water trawling is so efficient that in the last 10 years numbers have fallen drastically. A survey found that the average length of the fish has decreased by 10 cm since 1990. This will have a harmful effect on the population and could ultimately lead to extinction.

One of the problems is improved technology which locates complete shoals of the fish. Trawl nets are dragged along the sea floor stirring up sediment and killing many delicate invertebrates. This will affect biodiversity. So many countries fish the waters that a political agreement must be found.

Extract from *Action Ecology* 2000

Practice examination questions

(a) Suggest two ways in which trawling is considered 'efficient'. (line 3) [2]

(b) Why was the decrease in the average length of the cod considered to have a 'harmful effect' on the population of the species? (line 6) [2]

(c) How may trawling result in reduced biodiversity? (line 13) [3]

(d) Suggest how 'political agreement' may help to increase breeding stocks. (line 14) [2]

4 Students wished to investigate whether a pond in a field was being affected by a nearby cattle feeding station.

They sampled the water in the pond using a net and caught the following organisms:

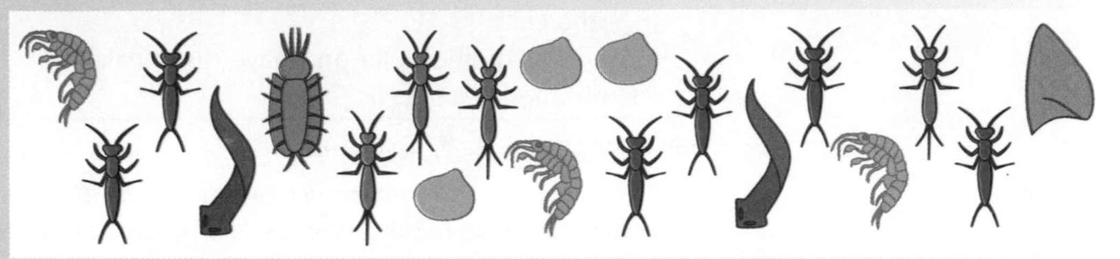

(a) Use the student' results to produce an index of diversity for the sample. [2]

(b) The students sampled another pond in the same way.
This pond was further from the feeding station and their results produced an index of diversity of 6.90.
What conclusions can be drawn from the students' results? [2]

Chapter 8

Human health and disease

The following topics are covered in this chapter:

- Health and lifestyle
- Disease
- Immunity
- Assisting immunity

8.1 Health and lifestyle

LEARNING SUMMARY

After studying this section you should be able to:

- define health
- understand the link between diet and good health
- describe ways in which food technology is used to increase food supplies
- understand how coronary heart disease can be affected by lifestyle
- describe the specific effects of smoking tobacco

How can we achieve good health?

AQA A 1.1
OCR 2.2.1

Good health is not just an absence of disease or infirmity. It is the physical, mental and social well-being of a person. The development of a healthy person begins in the uterus. It is important that the mother supplies the fetus with suitable nutrients for development, e.g. amino acids for proteins essential for healthy growth. The mother's diet is important for both her and the fetus.

Following birth, the emotional and social development are equally as important as physical development.

If a person is healthy then they may expect the following:

- an absence of disease
- an absence of pain
- to be fit and have good muscle tone
- an absence of stress
- to get along with other people in society
- to have a long life expectancy.

The importance of diet

OCR 2.2.1

The human diet is vital to good health. A newly born baby needs to feed on mother's first milk **(colostrum)** which is rich in antibodies. This gives immunity against some diseases. It is important that each of the following food classes is included in a person's diet, in a suitable proportion.

Examiner's tip
It is likely that you will be supplied with data about dietary constituents. Be ready to analyse the data and apply the principles of a balanced diet. Try to remember the main function of each substance. Analyse the dietary reference values for food energy and nutrients in the UK. The values indicate amounts of individual food components required per day and those which should not be exceeded.

The balanced diet

- **Carbohydrates** – sugars and starch supply metabolic energy; cellulose (dietary fibre) stimulates peristalsis so that constipation is prevented.
  Source *– potato and bread*
- **Proteins** – supply metabolic energy and are needed in growth and repair. **All enzymes are proteins**. Very important!
  Source *– meat and nuts*
- **Fats and oils** – supply metabolic energy and are needed in cell membrane formation, as they help to make phospholipids.
  Source *– butter and cooking oil*
- **Vitamins** – organic substances needed in minute quantities to maintain health, e.g. vitamin A. This is essential to make the pigment in the rods of the retina.
  Source *(vitamin A) – butter and carrots*

- **Minerals** – inorganic ions needed for a number of important roles in the body, e.g. iron. This is essential for the production of haemoglobin and so is vital for oxygen transport.
  Source (iron) – red meat, spinach
- **Water** – makes up over 50% of the content of blood plasma. It is needed for many functions including as a solvent and cooling the body down.

Remember that carbohydrates, proteins, lipids, vitamins, minerals, water and dietary fibre are all essential.

If a person does not eat enough of any one constituent of their diet then there is a deficiency disease, e.g. a protein deficiency causes kwashiorkor. If a person eats too much carbohydrate and fats then obesity and cardiovascular problems can result. A balanced diet is vital!

KEY POINT

Consumption of an unbalanced diet can lead to malnutrition.

Daily energy requirement

Eating, then respiring carbohydrates, proteins, fats and oils, supplies the energy needed for good health. Energy content of food is usually measured in **kilojoules (kJ)**. A diet rich in carbohydrates and/or fats and oils which exceeds daily requirements results in **obesity**. Large quantities of fat are stored around the body resulting in cardiovascular problems. If the kilojoule intake is regularly less than the daily requirement then a condition known as **anorexia nervosa** can develop. People with this condition have an emotional disorder of which absence of appetite is a symptom.

Maintaining food supply

OCR 2.2.1

With the world's population continually growing, it is becoming increasingly difficult to produce enough food and to store it for long enough to supply everybody with a balanced diet. The science of **food technology** is using new developments to try and increase food supply.

Compare artificial selection with natural selection (see page 95).

The two processes have similarities but in natural selection it is change of the environment which is the selective agent.

Selective breeding

This is *selective* breeding to improve specific domesticated animals and crop plants. Important points are:

- people are the **selective agents** and choose the parent organisms which will breed
- the organisms are chosen because they have **desired characteristics**
- the aim is to incorporate the desired characteristics from both organisms in their offspring
- the offspring must be **assessed** to find out if they have the desired combination of improvements (there is **no guarantee** that a cross will be successful!)
- offspring which have suitable improvements are used for breeding, the others are deleted from the gene pool (not allowed to breed).

Can you suggest four excellent features offered by this new variety?

Most modern crops have been produced by artificial selection. The Brussels sprout variety below was produced in this way. Many trials were carried out before the new variety was offered for sale.

Brilliant NEW FOR 2001
F1 Hybrid A brand new early cropping variety which produces dense, dark green buttons of excellent quality in September and October. Suitable for a wide range of soil types it also has a high resistance to powdery mildew and ring spot. Good for freezing. **2152 pkt £2.10**

Fertilisers

It is important that crop plants have access to all the minerals they require to give a maximum yield. Farmers supply these minerals in fertilisers, usually in the form NPK (nitrogen, phosphates and potassium). By supplying them with these minerals, nitrogen is available to make protein, a key substance for growth. Phosphates help the production of DNA, RNA and ATP. Potassium helps with protein synthesis and chlorophyll production. Other minerals are also needed like iron and calcium. The more a plant grows, the more its biomass increases and usually the greater the surface area for light absorption. The amount of photosynthesis increases proportionally. If a farmer is to reach the maximum productivity of a crop, fertiliser is vital.

Pesticides

If pests such as aphids or caterpillars begin to damage crops, then both quality and yield are reduced. Farmers combat pests by using chemical pesticides. Chemicals used to kill insects are insecticides. **Contact insecticides** kill insects directly but **systemic insecticides** are absorbed into the cell sap. Any insect consuming part of the plant or sucking the sap then dies.

Antibiotics

As well as treating crop plants with chemicals to improve yield, farmers can treat their animals with antibiotics. This is often done on a blanket basis not as a response to diseases. In this way, farmers hope to prevent disease in their animals and increase growth rates.

Microorganisms for food

Microorganisms are being increasingly used as a source of food. They include bacteria, fungi, yeasts and algae. These foods are often referred to as **single cell protein**. They have a number of advantages over more traditional sources of protein.

- They can be grown on waste products such as molasses and whey.
- They have a high growth rate.
- They do not need large areas of land for their cultivation.
- They often contain more fibre and less saturated fat than meat.
- They can be eaten by vegetarians.

Try to avoid talking about cost unless you qualify the statement. Single cell protein production plants are expensive to set up but they do run on inexpensive wastes.

There are disadvantages, however. They may need flavourings to make them more palatable and some are unsuitable for human consumption because of their high RNA content.

Food preservation

In order to transport food long distances or store it for future consumption, methods of food preservation are used. This will prevent **microbial spoilage** by making the conditions unsuitable for microorganisms to grow:

- adding salt or sugar – this will reduce the water potential
- pickling – this reduces the pH
- freezing – this will reduce the temperature so that microbial enzymes are inactive
- heat treatment – pasteurisation to 70°C will kill harmful microorganisms and heating to higher temperatures will kill all microbes
- irradiation – exposure to gamma radiation will kill any microorganisms present and inhibit sprouting and ripening.

Coronary heart disease and lifestyle

AQA 1.5
EDEXCEL 1.1.10-15
OCR 2.2.1

Atherosclerosis is a major health problem caused by eating saturated fats. This circulatory disease may develop as follows:

- yellow fatty streaks develop under the lining of the **endothelium** on the inside of an artery
- the streaks develop into a fatty lump called an **atheroma**

- the atheroma is made from **cholesterol** (taken up in the diet as well as being made in the liver)
- dense **fibrous tissue** develops as the atheroma grows
- the endothelial lining can **split**, allowing blood to contact the fibrous atheroma
- the damage may lead to a blood clot and an artery can be blocked.

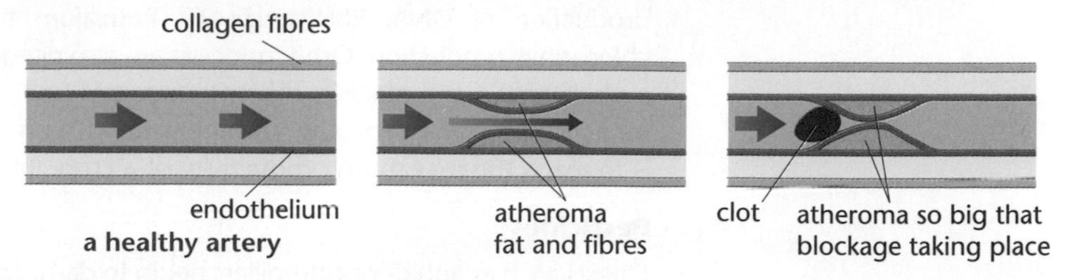

Remember that the clotting of blood can occur for other reasons. There may be damage at other positions around the body. Blockage of this type is **thrombosis.**

Increasing constriction of an artery caused by **atherosclerosis** and **blood clots** reduces blood flow and increases blood pressure. If the artery wall is considerably weakened then a bulge in the side appears, just like a weakened inner tube on a cycle tyre. There is a danger of bursting and the structure is known as an **aneurysm.**

It is possible for a blood clot formed at an atheroma to break away from its original position. It may completely block a smaller vessel, this is known as an **embolism.**

Coronary heart disease

If the artery which supplies the heart (coronary artery) is partially blocked, then there is a reduction in oxygen and nutrient supply to the heart itself. This is called **coronary heart disease** (CHD). The first sign is often **angina,** the main symptom being sharp chest pains. If total blockage occurs then **myocardial infarction** (heart attack) takes place.

Lifestyle and CHD

Many aspects of lifestyle influence the condition of the cardiovascular system. CHD is a multi-factorial disease and the risk of developing CHD depends on a number of factors.

Statistically people have a greater chance of avoiding CHD if they:

- consume a low amount of saturated fat in their diet
- do not drink alcohol excessively
- consume a low amount of salt
- do not smoke
- are not stressed most of the time
- exercise regularly.

New drugs called statins are being prescribed in an effort to reduce CHD. They are thought to work by decreasing LDL levels in the blood.

Saturated fats (see lipid structure pages 24–25) are found in large quantities in animal tissues. Eating large quantities of saturated fats seems to increase the risk of CHD. It seems that saturated fat increases the levels of **low-density lipoproteins** (LDLs) in the blood. LDLs transport cholesterol in the blood and high levels of LDLs seem to increase the risk of atheroma formation. Other fats such as polyunsaturated fats seem to increase the levels of **high-density lipoproteins** (HDLs). This seems to give some protection against CHD.

Exercise has a **protective effect** on the **heart** and **circulation**. Activities such as jogging, walking, swimming and cycling can:

- reduce the resting heart rate
- increase the strength of contraction of the heart muscle
- increase the stroke volume of the heart (the volume of blood which is propelled during the contractions of the ventricles).

High salt levels have been shown to increase blood pressure, therefore increasing the risk of damage to atheromas.

Smoking also increases blood pressure and makes the blood more likely to clot.

Progress check

(a) Name a specific substance in food which can result in atherosclerosis.
(b) Describe and explain the structural changes which take place in a blood vessel as atherosclerosis develops.
(c) How can the damage caused by an atheroma result in a heart attack?

(a) saturated fat
(b) yellow fatty streaks develop under the cells lining the inside of a blood vessel, the streaks develop into a lump known as an atheroma, the atheroma is made of cholesterol, dense fibrous tissue develops and the lining of the vessel can split.
(c) a blood clot forms which can block the blood vessel completely. Prevention of oxygen supply to the heart results in myocardial infarction (heart attack).

What are the dangers of smoking tobacco?

AQA 1.4
OCR 2.2.2

Each person has another choice to make, to smoke or not to smoke. The government health warning on every cigarette packet informs of health dangers but many young people go ahead and ignore the information.

Effects of tobacco smoking

- **Nicotine** is the active component in tobacco which **addicts** people to the habit.
- **Tars** coat the alveoli which **slows down exchange** of carbon dioxide and oxygen. If less oxygen is absorbed then the smokers will be less active than their true potential.
- **Cilia** lining bronchial tubes are coated then **destroyed,** reducing the efficiency in getting rid of pollutants which enter the lungs. These pollutants include the cigarette chemicals themselves.
- **Carbon monoxide** from the cigarette gases **combines with haemoglobin** of red blood cells rather than oxygen. This **reduces oxygen transport** and the smokers become less active than their potential. Ultimately it may lead to heart disease.
- The **bronchi** and **bronchioles** become **inflamed,** a condition known as **bronchitis**. This causes irritating fluid in the lungs, **coughing** and increased risk of heart disease. A number of bronchitis sufferers die each year.
- The walls of the alveoli break down reducing the surface area for gaseous exchange. Less oxygen can be absorbed by the lungs, leaving the **emphysema** sufferers extremely breathless. They increase their breathing rate to compensate but still cannot take in enough oxygen for a healthy life. Chronic emphysema sufferers need an oxygen cylinder to prolong their life.
- **Blood vessel elasticity is reduced** so that serious damage may occur. Ultimately a heart attack can follow.
- The **carcinogens** (cancer-causing chemicals) of the tobacco can result in **lung cancer. Malignant growths** in the lungs develop uncontrollably and cancers may spread to other parts of the body. Death often follows. Smokers have a greater risk of developing other cancers than non-smokers, e.g. more smokers develop cervical cancer.

Learn the characteristics of each disease carefully. There are so many consequences of smoking that you may well mix them up.

Even non-smokers can develop any of the above symptoms, but the probability of developing them is increased by smoking. Being in a smoky atmosphere each day also increases the chances.

The role of statistics

The government health warning on cigarette packets informs people of the risks of smoking. The WHO (World Health Organisation), governments and local authorities have collected statistics on many diseases over the years. These are used in education packs and posters to warn of risk factors. People can take precautions and use the information to avoid health dangers and take advantage of vaccination programmes.

Where education is not successful then related diseases follow. This acts as a drain on the National Health Service. Many operations which would have been unnecessary are performed to save people's lives, e.g. where coronary blood vessels are dangerously diseased a by-pass operation is the answer. The ideal situation is that education is successful, but realistically the aim is to balance prevention and cure.

8.2 Disease

After studying this section you should be able to:

- *define disease*
- *recall the causes, symptoms and control of a range of diseases including cholera, tuberculosis, malaria and AIDS*

LEARNING SUMMARY

What is a disease?

AQA 1.1
OCR 2.2.2

A disease is a **disorder** of a tissue, organ or system of an organism. As a result of a disorder, **symptoms** are evident. Such symptoms could be the failure to produce a particular digestive enzyme, or a growth of cells in the wrong place. Normal bodily processes may be disrupted, e.g. efficient oxygen transport is impeded by the malarial parasite, *Plasmodium*.

Different types of disease

Infectious disease by pathogens

Pathogens attack an organism and can be passed from one person to another. Many pathogens are spread by a **vector** which carries them from one organism to another without being affected itself by the disease. Pathogens include bacteria, viruses, protozoa, fungi, parasites and worms.

Most exam candidates recall that pathogens are responsible for disease. However, there are more causes of disease! If a question asks for different types of disease then giving a range of pathogens will not score many marks. Give genetic diseases, etc.

Genetic diseases

These can be passed from parent to offspring, e.g. haemophilia and cystic fibrosis.

Dietary related diseases

These are caused by the foods that we eat. Too much or too little food may cause disorders, e.g. obesity or anorexia nervosa. Lack of vitamin D causes the bone disease rickets, the symptoms of which are soft weak bones which bend under the body weight.

Environmentally related diseases

Some aspects of the environment disrupt bodily processes, e.g. as a result of nuclear radiation leakage, cancer may develop.

An **auto-immune disease** may be **environmentally** caused, e.g. the form of leukaemia where phagocytes destroy a person's red blood cells may be caused by radiation leakage.

Auto-immune disease

The body in some way attacks its own cells so that processes fail to function effectively.

How are infectious diseases transmitted?

The pathogens which cause infectious diseases are transmitted in a range of ways.

- Direct contact – sexual intercourse enables the transmission of syphillis bacteria; a person's foot which touches a damp floor at the swimming baths can transfer the Athlete's foot fungus.

Be prepared to answer questions about diseases not on your syllabus. The examiners will give data and other information which you will need to interpret. Use your knowledge of the principles of disease transmission, infection, symptoms and cure.

- Droplet infection – a sneeze propels tiny droplets of nasal mucus carrying viruses such as those causing influenza.
- Via a vector – if a person with typhoid bacteria in the gut handles food the bacteria can be passed to a susceptible person.
- Via food or water – chicken meat kept in warm conditions encourages the reproduction of *Salmonella* bacteria which are transferred to the human consumer, who has food poisoning as a result.
- Via blood transfusion – as a result of receiving infected blood a person can contract AIDS.

Some infectious diseases have serious consequences to human life. The incidence of infectious diseases may vary according to the climate of the country, the presence of vectors, the social behaviour of people and other factors.

Disease file – cholera

OCR 1.3

Cause of disease

Vibrio cholerae (bacteria) in the faeces or vomit of a human sufferer or human carrier which contaminate water supplies.

Transmission of microorganism

Contaminated water spreads the bacteria. Poor sanitary behaviour of people who are carriers and those who have contracted the disease are responsible. Faeces enters rivers which may be used for bathing, drinking, or irrigation. The bacteria survive outside the human body for around 24 hours. They can also contaminate vegetables and can be passed to a person in this way.

Outline of the course of the disease and symptoms

The bacteria reach the intestines where they breed. They secrete a toxin which increases secretion of chloride ions into the intestine, giving severe diarrhoea. Death is a regular consequence, due to dehydration, but some people do recover.

Prevention

Education about cleanliness and sewage treatment. Good sanitation is vital. Suitable treatment of water to be consumed by people, e.g. chlorination which kills the bacteria. Use of disinfectant also kills the bacteria. Early identification of an outbreak followed by control.

Cure

Tetracycline antibiotics kill organisms in the bowel. Immunisation is not very effective. It will help some individuals but not stop them from being carriers, so epidemics are still likely.

Oral rehydration therapy is usually used to combat the dehydration caused by diarrhoea and vomiting.

Simply drinking water is ineffective because the electrolyte losses also need replacing. The standard treatment is to restore fluids intravenously with water and salts. This requires trained personnel and materials which are not always available.

However, it was discovered that the body can absorb a simple solution containing both **sugar and salt**. The dry ingredients can be mixed and packaged, and then the solution can be given by people with minimal training. In cholera, the toxin interferes with one transport system in the intestine lining but the sodium-glucose co-transporter still operates. This allows absorption of glucose, sodium and water.

Oral rehydration can therefore be accomplished by drinking frequent small amounts of an **oral rehydration solutions**.

Disease file – tuberculosis

OCR 2.2.2

Cause of disease

Mycobacterium tuberculosis (bacterium) via droplet infection.

Transmission of microorganism

Coughs and sneezes of sufferers spread tiny droplets of moisture containing the pathogenic bacteria. People then inhale these droplets and may contract the disease.

Outline of the course of the disease and symptoms

The initial attack takes place in the lungs. The alveoli surfaces and capillaries are vulnerable and lesions occur. Some epithelial tissues begin to grow in number but these cannot carry out gaseous exchange. Inflammation occurs which stimulates painful coughing. Intense coughing takes place which can cause bleeding. There is much weight loss. Weak groups of people, like the elderly, or those underweight are more prone to the disease.

Prevention

Mycobacterium bovis causes tuberculosis in cattle. It can be passed to humans via milk. It causes an intestinal complaint in humans. It is important that cows are kept free of *M. bovis* by antibiotics.

The BCG vaccination is the injection of a weakened form of this microbe. This vaccination stimulates antibodies which are effective against both *M. tuberculosis* and *M. bovis*.

Mass screening using **X-rays** can identify 'shadows' in those people with scar tissue in the lungs.

Sputum testing identifies the presence of the bacteria in sufferers. Sufferers can be treated with antibiotics. Once cured they cannot pass on the pathogen so an epidemic may be prevented.

Skin testing is used. Antigens from dead *Mycobacteria* are injected just beneath the skin. If a person has been previously exposed to the organism then the skin swells which shows that they already have resistance, i.e. they have antibodies already. Anyone whose skin does not swell up is given the **BCG vaccination**. This contains attenuated *Mycobacterium bovis* to stimulate the production of antibodies against both *M. bovis* and *M. tuberculosis*.

Cure

Use of antibiotics such as streptomycin.

Disease file – malaria

OCR 2.2.2

Cause of disease

There are many variants of the malarial parasite, *Plasmodium* (protozoa).

Transmission of the microorganism

The vector which carries the *Plasmodium* is a female *Anopheles* mosquito. The mosquito feeds on a mammal which may be suffering from malaria. It does this at night by inserting its 'syringe-like' stylet into a blood vessel beneath the skin. The mosquito feeds on blood and digests the red blood cells which releases the malarial parasites. These burrow into the insect's stomach wall where they breed; some then move to the salivary glands. Next time the mosquito feeds it secretes saliva to prevent clotting of the blood. This secretion introduces the parasites into the person's blood, who is likely to contract the disease.

Outline of the course of the disease and symptoms

After entry into the blood, **sporozoites** invade the **liver** releasing many **merozoites**. Each merozoite infects a **red blood cell** producing even more merozoites. Millions of these parasites are released into the blood causing a fever. As a result, the sufferer develops a range of symptoms including pains, exhaustion, aching, feeling cold, sweating and fever. The increased body temperature attracts mosquitoes even more, so a person with malaria acts as a reservoir for parasites.

Prevention

The most effective methods of prevention are those which **destroy the vector.** Spraying **insecticide** onto lake surfaces kills mosquito larvae.

Oil poured on lake surfaces prevents air entering the breathing tubes of the mosquitoes, so they die. Fish can be introduced into lakes as **predators** to eat the larvae. This is an example of **biological control**.

Combinations of these tests are used in different countries. Where there are outbreaks of the disease the systems are activated.

Sometimes ponds are **drained** to remove the mosquitoes' breeding area. People in areas where malaria is endemic **cover up** all waste **tin cans** and **plastic containers.** If they were to fill up with rain water then the mosquitoes have another habitat to breed in. The bacterium, ***Bacillus thuringiensis*** is used to destroy mosquitoes. Mosquito nets exclude mosquitoes from buildings and are even used over beds. **Electronic insect killer** techniques can be used which attract the mosquito via ultra-violet light then kill them by application of voltage. **Drugs** are used so that even if a person is bitten by a mosquito any *Plasmodia* entering the blood fail to develop further.

Cure

It is necessary to isolate and treat the sufferer. This also reduces the spread of the disease. Drugs are used to kill the parasites in the blood and reduce the symptoms. People are constantly attempting to find different ways of preventing this killer disease.

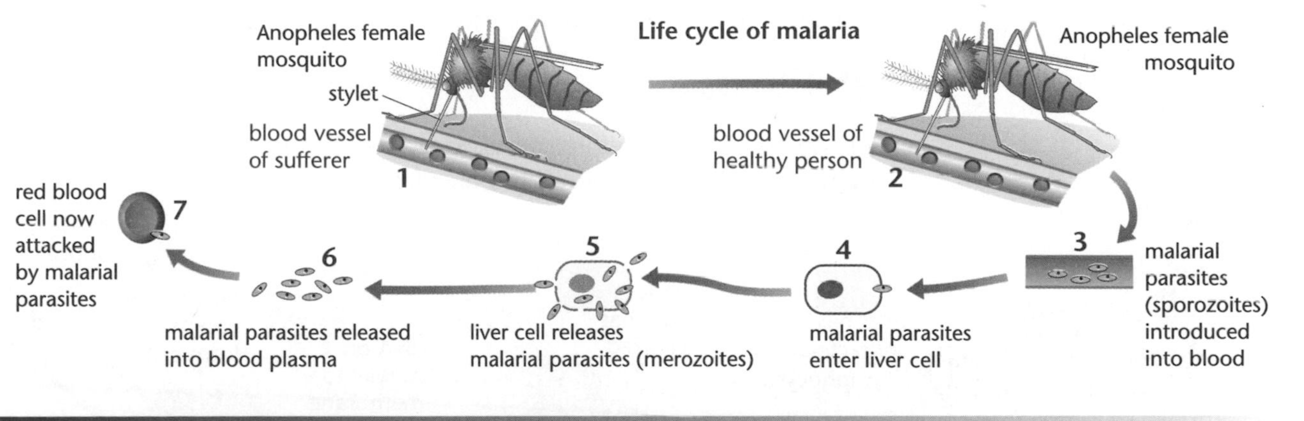

Progress check

A mosquito carries the malarial parasite, *Plasmodium*. The female mosquito feeds on a mammal by inserting its 'syringe-like' stylet into a blood vessel beneath the skin. The mosquito feeds on blood and digests the red blood cells releasing the malarial parasites. These burrow into the insect's stomach wall and breed there, then some move to the salivary glands. Next time the mosquito feeds it secretes saliva. The saliva introduces the parasites into the person's blood.

(a) (i) Which species of mosquito transmits malaria?
(ii) Which organism causes the disease, malaria?
(iii) Does every mosquito bite transmit malaria?

Give a reason for your answer.

(b) Suggest how to reduce the spread of malaria.

(a) (i) *Anopheles* (ii) plasmodium (iii) no – mosquito must feed on sufferer first.

(b) Drain ponds where the mosquitoes breed; kill the mosquitoes with insecticide; pour oil on ponds to kill the larvae; introduce insectivorous fish as a form of biological control; use drugs such as chloroquine to cure people suffering from the disease; isolate people suffering from the disease; spray mosquitoes with a suspension of *Bacillus thuringiensis*; use of preventative anti-malarial drugs.

Disease file – AIDS (Acquired Immune Deficiency Syndrome)

OCR 2.2.2

Cause of disease

This is caused by HIV (human immune deficiency virus). It is a retrovirus, which is able to make DNA with the help of its own core of RNA.

Transmission of microorganism

This takes place by the exchange of body fluids, transfusion of contaminated blood, or via syringe needle 'sharing' in drug practices.

Outline of the course of the disease and symptoms

Scientists are constantly trying to find a **cure**. None has been found yet.

Destruction of T-lymphocyte cells

The HIV protein coat attaches to protein in the plasma membrane of a T-lymphocyte. The virus protein coat fuses with the cell membrane releasing RNA and reverse transcriptase into the cell. This enzyme causes the cell to produce DNA from the viral RNA. This DNA enters the nucleus of the T-lymphocyte and is incorporated into the host cell chromosomes. The gene representing the HIV virus is permanently in the nucleus from now on and can be dormant for years. It may become activated by an infection. Viral protein and viral RNA are made as a result of the infection.

Many RNA viral cores now leave the cell and protein coats are assembled from degenerating plasma membranes. Other T-lymphocytes are attacked. Cells of the lymph nodes and spleen are also destroyed. Viruses appear in the blood, tears, saliva, semen and vaginal fluids. The immune system becomes so weak that many diseases can now successfully invade the weakened body.

Prevention

Screening of blood before transfusions. Use of condoms and remaining with one partner. No use of contaminated needles.

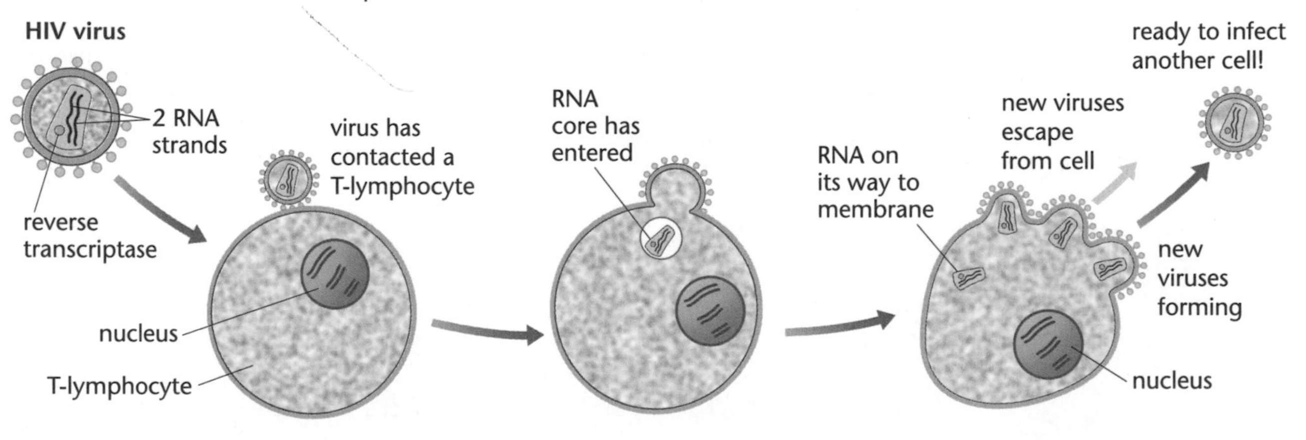

8.3 Immunity

After studying this section you should be able to:

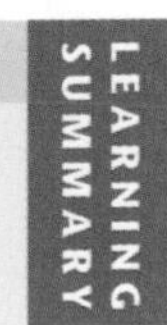

- *describe the mechanisms used by the body to try and prevent pathogen entry*
- *describe and explain the action of the body's immune system*

Survival against the attack of pathogens

OCR 2.2.2

Many pathogenic organisms attack people. They are not all successful in causing disease. We have immunity to a disease when we are able to resist infection. The body has a range of ways to prevent the disease-causing organism from becoming established.

- A tough protein called **keratin** helps skin cells to be a formidable **barrier** to prevent pathogens entering the body.
- An enzyme, **lysozyme**, destroys some microorganisms and can be found in sebum, tears and saliva.
- **Hydrochloric acid** in the stomach kills some microorganisms.
- The bronchial tubes of the lungs are lined with **cilia**. Microorganisms which enter the respiratory system are often trapped in mucus which is then moved to the oesophagus. From here they move to the stomach where many are destroyed by hydrochloric acid or digested.
- **Blood clotting** in response to external damage prevents entry of microorganisms from the external environment.

The methods the body uses to prevent microorganisms entering the bloodstream are sometimes unsuccessful. When the microorganisms invade and then breed in high numbers, we develop the **symptoms**. **White blood cells** enable us to destroy invading microorganisms. They may destroy the microorganisms quickly before they have any chance of becoming established, so the person would not develop any symptoms. Sometimes there are so many microorganisms attacking that the white blood cells cannot destroy all of them. Once the pathogens are established the symptoms of a disease follow, but for most diseases, after some time, the white blood cells eventually overcome the disease-causing organisms.

The roles of the white blood cells (leucocytes)

AQA 1.6
OCR 2.2.2

There are a number of different types of **leucocytes**. Most are produced from **stem cells** in the **bone marrow**. Different stem cells follow alternative maturation procedures to produce a range of leucocytes. Leucocytes have the ability to recognise self chemicals and non-self chemicals. Only where non-self chemicals are recognised will a leucocyte respond. Proteins and polysaccharides are typical of the complex molecules which can trigger an immune response.

Phagocytes

Phagocytes can move to a site of infection through capillaries, tissue fluid and lymph as well as being found in the plasma. They move towards pathogens which they destroy by the process of phagocytosis. This is often called engulfment and involves the surrounding of a pathogen by pseudopodia to form a food vacuole. Hydrolytic enzymes from lysosomes complete the destruction of the pathogen.

Neutrophils are one type of phagocyte. Proteins in plasma called **opsonins** attach to a pathogen. These opsonins enable the phagocyte to engulf the pathogen.

Macrophages are another type of phagocyte which work alongside T-lymphocytes.

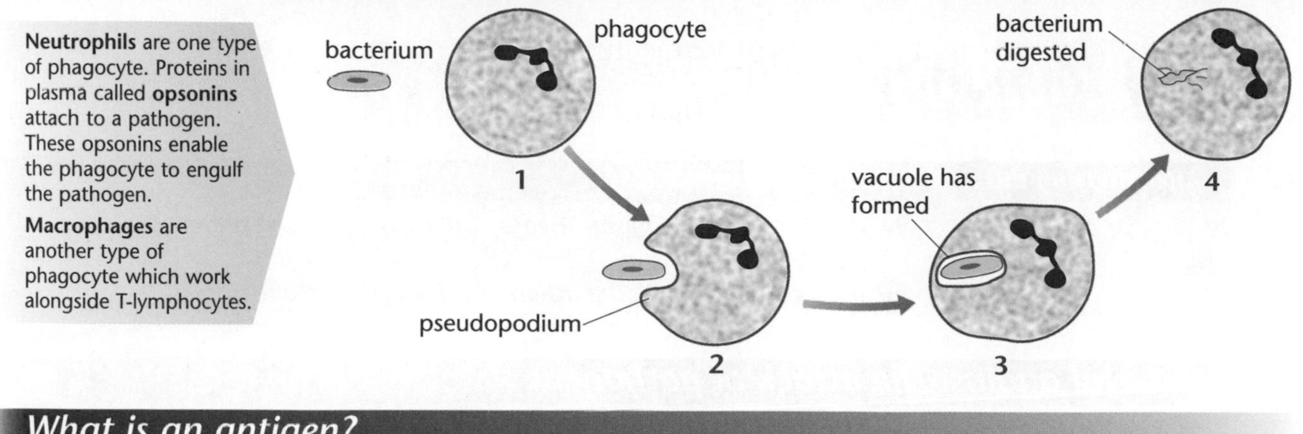

What is an antigen?

AQA A 1.6
OCR 2.2.2

As an individual grows and develops, complex substances such as proteins and polysaccharides are used to form cellular structures. Leucocytes identify these substances in the body as 'self' substances. They are ignored as the leucocytes encounter them daily. 'Non-self substances', e.g. foreign proteins which enter the body, are immediately identified as 'non-self'. These are known as **antigens** and trigger an immune response.

Lymphocytes

White blood cells (leucocytes) constantly check out proteins around the body. Foreign protein is identified and attack is stimulated.

In addition to phagocytes, there are other leucocytes called lymphocytes. There are two types of lymphocyte, **B-lymphocytes** and **T-lymphocytes**.

B-lymphocytes begin development and mature in the bone marrow. They produce antibodies, known as the **humoral response**.

T-lymphocytes work alongside phagocytes known as macrophages; this is known as the **cell-mediated response**. A macrophage engulfs an antigen. This antigen remains on the surface of the macrophage. T-lymphocytes respond to the antigen, dividing by mitosis to form a range of different types of T-lymphocyte cells.

- **Killer T-lymphocytes** adhere to the pathogen, secrete a toxin and destroy it.
- **Helper T-lymphocytes** stimulate the production of antibodies.
- **Suppressor T-lymphocytes** are inhibitors of the T-lymphocytes and plasma cells. Just weeks after the initial infection, they shut down the immune response when it is no longer needed.
- **Memory T-lymphocytes** respond to an antigen previously experienced. They are able to destroy the same pathogen before symptoms appear.

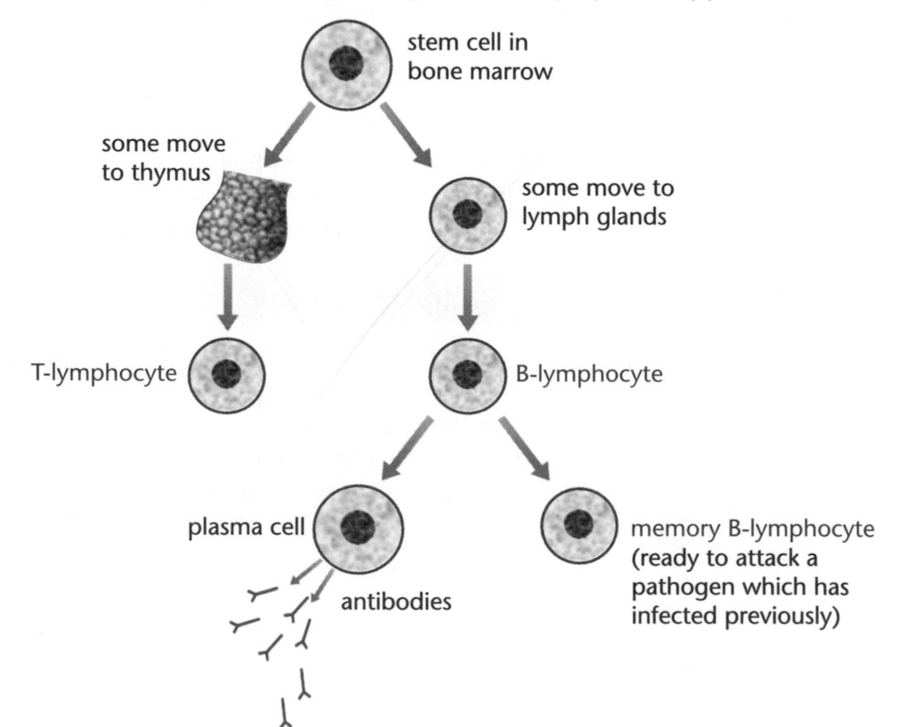

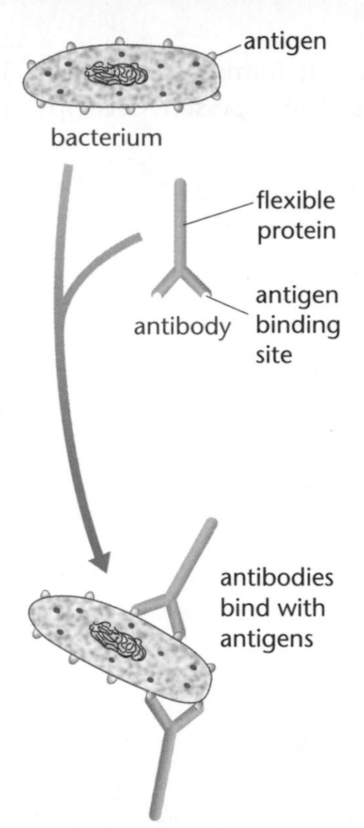

How do antibodies destroy pathogens?

The diagram on the left shows antibodies binding to antigens. The descriptions below show what can happen immediately after the binding takes place.

There are three main ways in which antibodies destroy pathogens.

- **Precipitation**, by linking many antigens together. This enables the phagocytes to engulf them.
- **Lysis**, where the cell membrane breaks open, killing the cell.
- **Neutralisation** of a chemical released by the pathogen, so that the chemical is no longer toxic.

8.4 Assisting immunity

After studying this section you should be able to:

- *describe the origin of different types of immunity*
- *explain how immunity can be stimulated by vaccinations*

LEARNING SUMMARY

Stimulating the immune system

AQA 1.6
OCR 2.2.2

Newborn babies are naturally protected against many diseases such as measles and poliomyelitis. This is because they have received antibodies from their mother in two ways:

- across the placenta
- via breast feeding.

This is one major advantage of breast feeding over bottle feeding of babies.

This type of immunity is called **passive natural immunity**. It is possible to give people injections of antibodies that have been made by another person or animal. This type of immunity is called **passive artificial** immunity.

> Passive immunity does not last very long. This is because the antibodies do not persist in the blood for very long.
>
> KEY POINT

Active immunity

Once a person has recovered from certain diseases, e.g. measles, they rarely contract that disease again. This is because the memory cells formed from B-lymphocytes can survive in the blood for many years. If the antigen reappears, they can rapidly produce a clone of antibody producing cells. This **secondary response** is rapid and larger than the **primary response** and the antigen is rapidly destroyed.

This type of immunity is called **active natural immunity**.

Vaccines consist of dead or weakened (**attenuated**) forms of pathogens. They stimulate the production of memory cells so that the person develops active artificial **immunity**.

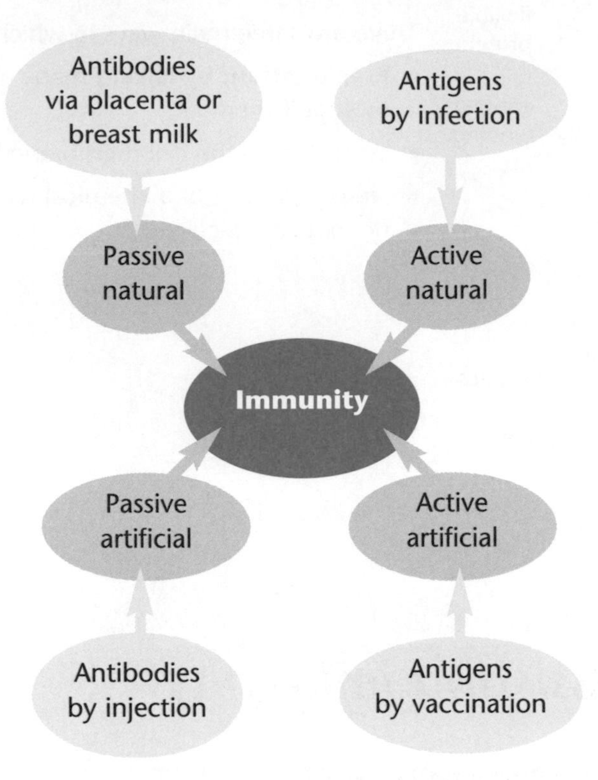

Unfortunately some pathogens show **antigenic variation**. This change in antigens renders vaccinations ineffective after a while and allows diseases such as influenza to strike many times in slightly different forms.

Scientists are now using antibody producing cells that have been fused with tumour cells. This produces cells that make large quantities of one type of antibody. They are called monoclonal antibodies. Because of their ability to target particular antigens, they might be useful in delivering drugs straight to target cells.

er 3 Enzymes

k and key – the substrate is a similar shape to
e active site; it fits in and binds with the active
e like a key (substrate) fitting into a lock (active
e); induced fit – the substrate is not a matching
for the active site, but as the substrate
proaches, the active site changes into an
propriate shape.
ersible – enters active site but will come out again
versible – binds permanently with enzymes [4]
[Total: 4]
not contaminate the product; they can be
in and again. [2]
[Total: 2]
ertiary structure of the enzyme is responsible
e further folding of the protein;
ves the shape of the active site;
ive site in amylase is specific to starch, lipid is
to bind to the active site of amylase,
re amylase cannot break down lipid. [3]
[Total: 3]
[1]
ostable enzymes are effective at high
eratures [1]
[Total: 2]

5 (a) The substrate molecule collides with the active site of the enzyme; as it approaches, the active site changes shape to become compatible with the substrate shape. [2]
(b) Non-competitive inhibitor molecule binds with part of enzyme other than the active site; as a result the active site changes shape; so the substrate can no longer bind with the active site. [3]
[Total: 5]

6 (a) NH_2 (amino group) [1]
(b) The enzymes then have more ends to attack. [2]
[Total: 3]

4 Exchange

change in size because the water potential
de the cell equals the water
ntial of the solution outside the cell. [1]
water potential of the solution outside the
s more negative than the water potential
e the cell. [1]
water potential of the solution outside the
s less negative than the water potential
e the cell. [1]
[1]
[Total: 4]
h use a protein carrier molecule. [1]
nsport needs energy or mitochondria,
acilitated diffusion does not.
transport allows molecules to move from
oncentrated solution to a higher
ted solution.
transport allows molecules to move
concentration gradient. [1]
[Total: 2]
in blade allows maximum light
on;
as a waxy cuticle to reflect excess
allow entry of enough light for
thesis;
phyll cells have many chloroplasts to
e maximum amount of light;
de cells pack closely together to absorb
num amount of light;

- guard cells open stomata to allow carbon dioxide in and oxygen out during photosynthesis;
- air spaces in the mesophyll store lots of carbon dioxide for photosynthesis;
- xylem of the vascular bundles bring water to the leaf for photosynthesis;
- phloem takes the carbohydrate away from the leaf after photosynthesis.

[any 6 points] [6]
[Total: 6]

4 (a) alveoli have a very high surface area; they are very close to many capillaries; capillaries are one cell thick/very thin/have squamous epithelia; they are kept damp which facilitates diffusion. [4]
(b)
- there are many gill filaments, which give a very high surface area;
- the gill filaments are very thin;
- they have many capillaries;
- they have gill plates which increase the surface area further;
- the blood and water directions are opposite which maximises diffusion/they use countercurrent flow to maximise diffusion.

[any 4 points] [4]
[Total: 8]

Sample question and model answers

The graph below shows the relative numbers of antibodies in a person's blood after the vaccination of attenuated viruses. Vaccinations were given on day 1 then 200 days later.

In examinations you are regularly given graphs. Make sure that you can link the idea being tested. This should help you recall all of the important concepts needed. All you need to do after this is **apply** your knowledge to the given data.

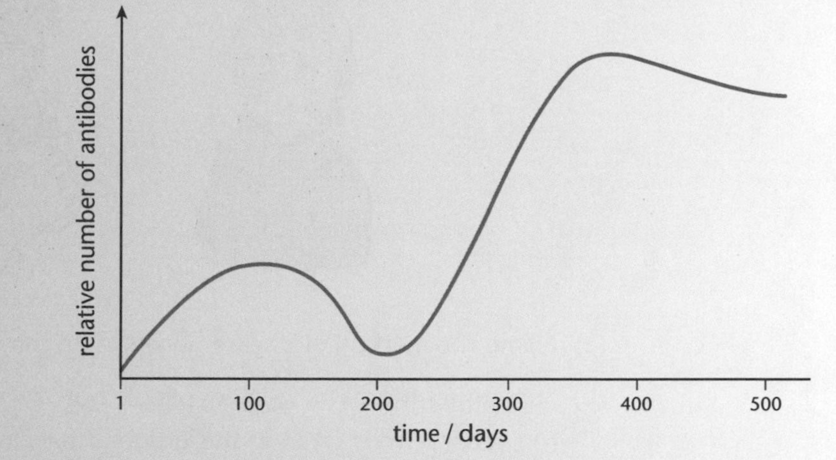

(a) Why is it important that viruses used in vaccinations are attenuated? [1]

If they were active then the person would contract the disease.

(b) Suggest **two** advantages of giving the second vaccination. [2]

A greater number of antibodies were produced.
The antibodies remain for much longer after the second vaccination.

(c) Which cells produced the antibodies during the primary response? [1]

B-lymphocytes.

(d) Why was there no delay in the secondary response to vaccination? [3]

Because the first vaccination had already been given, memory B-lymphocytes had been produced which respond to the viruses more quickly.

(e) Describe how a virus stimulates the production of antibodies? [3]

Antigen in the protein 'coat' or capsomere stimulate the B-lymphocytes.

If you gave B-lymphocytes as a response it would be wrong! B-lymphocytes secrete antibodies.

(f) Apart from producing antibodies, outline FOUR different ways that the body uses to destroy microorganisms. [4]

Phagocytes by engulfment; T-lymphocytes attach to microorganisms and destroy them; hydrochloric acid in the stomach; lysozyme in tears.

Practice examination questions

1 The illustration shows how the bacterium which causes TB can be transmitted from one person to another.

(a) Name the method of disease transmission shown in the illustration. [1]

(b) Sometimes the bacteria infect people but they do not develop symptoms.

(i) What term is given to this group of people? [1]

(ii) Explain why these people may be a greater danger to a community than those who actually suffer from the disease. [2]

(c) (i) What can be given to a person infected with TB to help destroy the bacteria? [1]

(ii) Explain the role of each of the following in destroying TB bacteria.

Phagocyte
B-lymphocyte
T-lymphocyte [6]

2 The diagram shows the response of B-lymphocytes to a specific antigen.

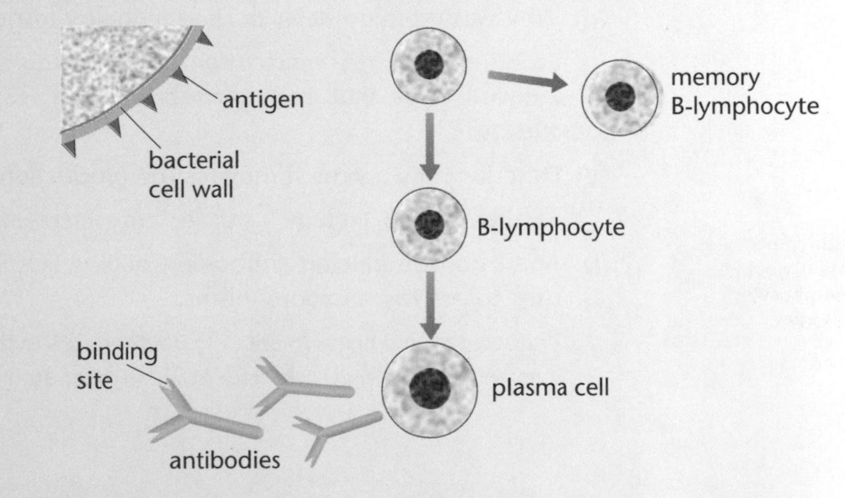

(a) (i) A plasma cell is bigger than a B-lymphocyte. Suggest an advantage of this. [1]

(ii) Describe the precise role of antibodies in the immune response. [3]

(iii) What is the advantage of memory B-lymphocytes? [2]

(b) What is an auto-immune disease? Give an example. [2]

(c) A person contracts the virus which causes the common cold. Suggest why their lymphocytes may fail to destroy the pathogen. [1]

Practice examination answe

Chapter 1 Biological molecules

1 (a) $\frac{90}{100} = 0.9$ [2]

(b) X [1]

(c) The solvent front would have reached the edge of the paper. [1]

[Total: 4]

2 (a)

RCOOH $HOCH_2$
RCOOH + HOCH
RCOOH $HOCH_2$
fatty acids glycerol [2]

(b) Emulsion test: add the sample to ethanol and mix; decant or pour into water; if a fat is present a white emulsion forms on the surface. [3]

[Total: 5]

3 (a) peptide bond/

(b) –COOH/carbo

(c) primary struct

4 (a) The latent he energy is nee much body

(b) High specifi needs a lot significantly overheat ea

(c) Cohesive f xylem.

Chapter 2 Cells

1 (a) phospholipid [1]

(b) (i) Substance approaches a carrier protein molecule; carrier protein activated by ATP; protein changes shape allowing the substance into the cell.

(ii) Substance approaches a carrier protein; this may be a channel protein and substances pass through without any ATP necessary. [4]

[Total: 5]

2 (a) C = Golgi body
B = Centrioles
A = Cell membrane
E = Mitochondria
D = Rough endoplasmic reticulum [5]

(b) Correct measure of width nucleus (~2.0 cm) convert to micrometres (20 000 μm) divide size by 5 000 (4 μm) [3]

(c) Liver cells carry out many functions; Need large amounts of energy/ATP [2]

[Total: 10]

3 (a) cell wall mitoch

(b) mitoch riboson [any tv

4 (a) lens X

(b) electr

(c) (i) a

(ii) i

Chapter 5 Transport

1 (a) to the body core [1]

(b) less blood reaches the superficial capillaries of the skin; so less heat is lost by conduction, convection and radiation; blood in the body core better insulated by the adipose layer of the skin [4]

[Total: 5]

2 (a) 60 x 120 x 5 = 36 000 ml / 36 litres [2]

(b) • blood is transported more quickly;
• more oxygen taken up at the lungs/more carbon dioxide excreted at the lungs;
• more oxygen reaches the muscles;
• more glucose reaches the muscles;
• so muscles contract more effectively.
[any 4 points] [4]

(c) (i) • slower breathing rate;
• more alveoli accessed for exchange;
• intercostal muscles more effective.
[any 2 points] [2]

(ii) • improved muscle tone;
• greater muscular strength;
• more capillaries in muscles. [any 2 points] [2]

[Total: 10]

3 (a) The amount of water lost by transpiration is exactly matched by the amount taken up by the leaf. [1]

(b) (i) xylem [1]

(ii) fill the potometer under water; operate the valve to get rid of air bubbles; when removing the leaf from the tree, the stalk or petiole must be put in water immediately. [2]

(c) Volume of water = πr^2 x 32 mm x 60
$= \frac{22}{7} \times 1 \times 1 \times 32 \times 60$
= 6034.3 mm^3 [3]

[Total: 7]

4 • extensive root system to absorb maximum water;
• large amount of water storage in leaves;
• thick cuticle;
• low numbers of stomata/sunken stomata;
• hairs to reduce turbulence. [any 3 points] [3]

[Total: 3]

5 (a) SAN/Sinoatrial node [1]

(b) (i) slows heart rate [1]

(ii) speeds up the rate [1]

(iii) speeds up the rate [1]

[Total: 4]

6 (a) (i) sieve tube [1]

(ii) companion cell has nucleus plus ribosomes; which make the proteins or enzymes; supplies enzymes to sieve tube via plasmodesmata. [3]

(b) (i) It is an active process; a pump is involved; phloem contents under high pressure. [3]

(ii) no starch in the phloem contents; did not change to blue-black; contained reducing sugar; did change to brick red. [4]

(iii) Hot wax kills phloem cells; so they cannot transport the radioactive carbohydrate; transport by phloem is an active process. [3]

[Total: 14]

Chapter 6 Genes and cell division

1 (a) metaphase [1]

(b) 4 [2]

[Total: 3]

2 (a) adenine and thymine are similar proportions because adenine binds with thymine; cytosine and guanine are similar proportions because cytosine binds with guanine [2]

(b) They should be identical in number but the scientists were operating at the limits of instrumentation. [1]

(c) Organic bases form the codes for different amino acids. Different sequences of amino acids form the different proteins specific to a species. [2]

[Total: 5]

3 (a) restriction endonuclease [1]

(b) Yes, the egg DNA shares several common bonds with the parents' DNA. [1]

(c) checking out who is the father of a child/paternity cases; crimes where blood samples or tissue or saliva is left and checked against suspects [2]

[Total: 4]

Chapter 7 Classification and biodiversity

1 A = Prokaryotae
B = Fungi
C = Protoctista
D = Plantae
E = Animalia [5]

[Total: 5]

2 (a)

	Mistle-thrush	*Song-thrush*
Kingdom	**Animalia**	**Animalia**
Phylum	Chordata	Chordata
Class	Aves	**Aves**
Order	Passeriformes	Passeriformes
Family	**Turdidae**	**Turdidae**
Genus	**Turdus**	**Turdus**
Species	**viscivorus**	**ericetorum**

[3]

(b) disruptive selection [1]

[Total: 4]

3 (a) electronic systems locate shoals of fish accurately; very large nets/small-mesh nets [2]

(b) many of the breeding size fish have already been caught; some fish may never reach breeding size as they are caught before they reach this size [2]

(c)
- trawling destroys some invertebrates;
- they may be the food of other organisms in a food web, so some animals may die out as a result;
- overfishing reduces fish numbers so that their predators may ultimately die out;
- nets catch other than target fish in the nets.

[any 3 points] [3]

(d) agree to quota numbers of fish; exclusion zones/exclusion times [2]

[Total: 9]

4 (a) N = 20

$$d = \frac{20 \times 19}{(3 \times 2) + (6 \times 5) + (2 \times 1) + (1 \times 0) + (4 \times 3) + (3 \times 2) + (1 \times 0)}$$

$$= \frac{380}{56} = 6.79$$

(b) Biodiversity was lower in pond near feeding station. Difference may not be significant.

Chapter 8 Human health and disease

1 (a) droplet infection [1]

(b) (i) carriers [1]

(ii) we do not know that they carry the pathogen as they display no symptoms, so the people do not avoid contact and pass on the bacteria [2]

(c) (i) antibiotics or named antibiotics [1]

(ii) **Phagocyte** – engulfs/produces pseudopodia/ phagocytosis; digests the bacterium/causes lysis of the bacterium [2]

B-lymphocyte – changes into plasma cell; makes antibodies [2]

T-lymphocyte – whole cell links to bacterial antigen sites; cell is usually destroyed by this; reacts to bacterial antigen [2]

[Total: 11]

2 (a) (i) ability to secrete more antibodies [1]

(ii) the antibodies have specific receptor sites which bind with the antigens; they have a flexible protein which changes angle to fit the antigens; antibodies result in the destruction of the antigen in some way/ neutralise toxin/cluster around antigens then cause precipitation/cause agglutination [3]

(iii) are produced when body first exposed to antigen; remain in body to react quickly when exposed to same antigen again [2]

(b) when the immune system attacks the person's own cells; pernicious anaemia/rheumatoid arthritis [2]

(c) the influenza virus often mutates so lymphocytes take longer to produce antibodies [1]

[Total: 9]

Index

Index

Notes